To Bob
with kind love
Ken

CW00551486

Basic Principles
of the
Finite Element Method

To Alice
my infinite element
in gratitute
for encouraging me to
persist in the completion of this book
and for loving companionship and support
over the last fifty years

Basic Principles
of the
Finite Element Method

K.M. Entwistle
Manchester Materials Science Centre
University of Manchester/UMIST

Book 711
First published in 1999 by
IOM Communications Ltd
1 Carlton House Terrace
London SW1Y 5DB

© IOM Communications Ltd

IOM Communications Ltd
is a wholly-owned subsidiary of
The Institute of Materials

ISBN 1 86125 084 3

Typeset, printed and bound in the UK by
Alden Press, Oxford

CONTENTS

PREFACE

The finite element method is now widely used in research in materials science and technology. For example it is the basis for the determination of the stress distribution in loaded specimens used in deformation and fracture studies. It is used to predict the mechanical behaviour of composite and of cellular solids and it is used to analyse materials processing of metals and polymers.

Materials science researchers use one of the many available commercial finite element packages to model problems in these areas. These materials scientists and technologists are not always well informed about the principles of the analytical methods that these packages use. One reason for this is that they find the existing texts difficult to read. There is an extensive list of finite element books written mostly for engineers or mathematicians. In them the authors make assumptions that the reader has a facility with matrix algebra, has a grounding in applied mechanics and has an awareness of energy principles that do not feature prominently in undergraduate materials science courses and, in consequence, are rarely the stock in trade of materials science researchers or technologists.

The objective of this book is to provide an introductory text which lays out the basic theory of the finite element method in a form that will be comprehensible to material scientists. It presents the basic ideas in a sequential and measured fashion, avoiding the use of specialist vocabulary that is not clearly defined. Occasionally some readers may find the detailed explanations to be over tedious, but I have deliberately erred on this side to avoid the greater criticism that essential steps that the reader needs are omitted. The basic principles are illustrated by a diversity of examples which serve to reinforce the particular aspects of the theory, and there are three finite element analyses which are presented in extenso with the detailed mathematics exposed. By this means some of the mystery that can envelop commercial finite element packages is penetrated.

Such is the extensive scale of finite element knowledge that any text of this introductory character must be selective in its choice of material. The criterion for the selection of topics has been guided by the wish to bring the readers to the point at the end of the book where they can develop their understanding further by reading the existing literature, in which there is a number of rigorous and scholarly texts with a wealth of detail on the more advanced aspects of the theory.

No list of recommended texts is included. The choice of texts is a matter of personal choice. The most fruitful way forward is to browse the library shelves or the bookshop to seek a text that addresses the area in which enlightenment is sought in a way which accords with the readers current knowledge. Mention must be made, however, to the work of Zienkiewicz. He and his school in Swansea made seminal advances in the theory of the finite element method and its application to engineering. His book with Taylor* on the finite element method must be regarded as the Authorised Version of the finite element bible.

I have deliberately used the second person plural in the text in order to emphasise my intention that the treatment of the subject should constitute an (inevitably) one-sided tutorial with the reader.

---

* Zienkiewicz O.C. and Taylor R.L. The Finite Element Method (in two volumes) 4th edition 1989 New York: McGraw-Hill

ACKNOWLEDGEMENTS

The author is grateful to a number of people for computational support. Professor Michael Burdekin's Structural Integrity Group in UMIST provided all the ABAQUS solutions. Dr. Sylvester Abuodha and Dr. Weiguang Xu were particularly helpful. Dr. Richard Day of the Manchester Materials Science Centre performed all the LUSAS calculations and gave general computational advice. Peter Entwistle of Frazer-Nash Consultancy Ltd., Dorking was a source of continuing assistance. It hardly needs to be said that the author takes full and complete responsibility for the way in which the data, generously provided by these supporters, are presented and interpreted.

Dr. Christopher Viney of the University of Oxford Department of Materials made valuable constructive comments on the manuscript.

CHAPTER 1   INTRODUCTION

We shall concern ourselves in this book with the application of the finite element method to the calculation of the elastic stress and strain distribution in loaded bodies. The method can be applied also to analyse problems involving electro-magnetic fields, acoustics, heat flow, soil mechanics and fluid mechanics, but we shall not address any of these areas in this book.

In this introduction we present an overview of the scope of the chapters that follow and, in the process, refer to specific terms in the finite element vocabulary which are of major importance and the explanation of which will occupy our energies in the ensuing text. These terms are dignified by capitals to give them prominence.

The basic idea in finite element elastic stress analysis is the following. The elastic body is first divided into discrete connected parts called FINITE ELEMENTS. These elements are commonly either triangles or quadrilaterals for two-dimensional models, with either straight or curved sides. For three-dimensional problems they might be tetrahedra or parallelepipeds. Examples are drawn in Fig. 1.1. The points at which the elements are connected together are called NODES, located at the corners of the elements and sometimes along the sides. The process of dividing the loaded body into elements is called (inelegantly) DISCRETEISATION. The pattern of elements is called a MESH. The selection of the mesh pattern and the element size is a matter of judgement and experience. Commercial finite element packages commonly have a programme that will form a mesh automatically.

The next step in the analysis is to establish a relation between the forces applied to the nodes of a single element and the nodal displacements produced by them. These two quantities are linked by the STIFFNESS MATRIX, through an equation of the form

$$(\text{nodal forces}) = [\text{stiffness matrix}] \times (\text{nodal displacements})$$

which is a particularly important feature of finite element analysis. We next combine together all the stiffness matrices of the individual elements into a single large matrix which is the GLOBAL STIFFNESS MATRIX for the whole body. This combination process is called ASSEMBLY.

It turns out that we will usually know the applied nodal forces and the unknowns are the nodal displacements. In order to arrive at the nodal displacements we need to INVERT the stiffness matrix equation to the form

$$(\text{nodal displacements}) = [\text{inverse stiffness matrix}] \times (\text{applied nodal forces}).$$

This matrix equation can be solved for the nodal displacements and an important part of the analysis has been encompassed. It is then possible to calculate the element stresses at selected points.

The process of setting up the stiffness matrix has two stages. The first involves relating the elastic strains in the element to the nodal displacements. We have first to make an assumption about the way in which the particle displacement at a point in the element varies with position. The simplest assumption is that the particle displacements vary linearly with position, which gives strains that are constant across the element. We shall analyse such a situation later when we use the constant strain triangular element. Higher order variations, for example parabolic or cubic, lead to strains that vary across the element and hold out the possibility of securing a closer approximation to the exact strain. The relations that define the law of variation of particle displacement with position in the element are called INTERPOLATION FUNCTIONS and are expressed in terms of what are called SHAPE FUNCTIONS. Once the set of shape functions has been selected, which will be related to the particular type of element chosen, it is a (tedious) formality to relate the elastic strains to the nodal displacements.

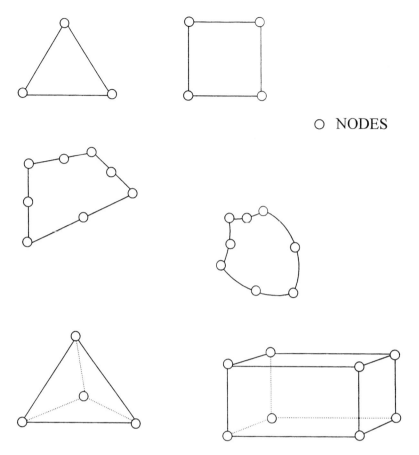

O NODES

Fig. 1.1 Examples of two-dimensional and three-dimensional finite elements showing the location of the nodes.

The second step in the generation of the stiffness matrix is to relate the nodal forces to the stresses in the element. This is not a straightforward task. It requires the use of energy principles in the form of either the VIRTUAL WORK PRINCIPLE or the MINIMISATION OF THE TOTAL POTENTIAL. These are alternatives. We favour the former but spend some time explaining both principles.

In the later stages of our development of the subject, we shall need to integrate functions over the whole area of an element. The functions that arise cannot be integrated analytically so we have to resort to numerical methods. The one we choose is GAUSSIAN QUADRATURE. This involves identifying special points in the element, called GAUSS POINTS or INTEGRATION POINTS at which the function to be integrated is evaluated. The integration is then accomplished by summing the product of the value of the function and what is called the WEIGHT-ING FACTOR at all the Gauss points. We explain the weighting factors and indicate how they are derived.

In finite element analysis we generate a large number of simultaneous equations. The most economical way to present these and to manipulate them is by using MATRIX REPRSENTA-TION. We devote a chapter to the presentation of some basic ideas about matrices and their processing.

We next look at the natural finite element systems that are assemblies of connected springs. Each spring is a finite element and the calculation of the spring displacements produced by a set of loads acting at the connecting points (the nodes) has some of the concepts and essence of finite element theory built into it. We carry out an analysis on a one-dimensional linear spring array to minimise the arithmetical complexity and bring out the basic ideas in as uncluttered a way as possible.

We then move on to analyse finite elements that are part of an elastic sheet or CONTINUUM. The first is a triangular constant strain element. The new feature here is the need to relate the nodal forces to the element stresses. In the case of the simple springs the spring tension is the equivalent to the stress and the relation between the end displacements and the spring tension is easily established. With the triangular element, however, arriving at the stresses within it from the nodal displacements is a more demanding task and we need to apply energy principles to that end. We devote a chapter to explaining these principles before embarking on the development of the relationships that we need to analyse finite element problems using triangular elements. There follows an example using only two triangular elements which makes it realistic to work through every stage in the calculations by hand and to obtain values for the element stresses for a prescribed loading system.

Higher order elements are then addressed in which the interpolation functions are parabolic not linear and the stress varies over the element. We concentrate on the 8-node quadrilateral element. The analysis that relates the element strains to the nodal displacements uses shape functions, and the integration that leads to the stiffness matrix uses Gaussian quadrature. Sections are devoted to explaining shape functions and numerical integration by Gauss quadrature.

An important feature of the analysis using higher order elements is a powerful concept of a generic MASTER ELEMENT. This is a simple square element from which it is possible to derive any real quadrilateral by a process called MAPPING. The nature of this process is explained. It is also explained that the basic analysis that leads to the stiffness matrix can be carried out on the generic master element and the results of this analysis can be applied to any real quadrilateral element derived from it by the mapping process.

After having established the relations needed to perform a finite element analysis with 8-node quadrilateral elements, we work through in detail an example comprising a single element in the form of a tapered sheet fixed at one edge and loaded at another. The first example uses four integration points. We then repeat the example using nine integration points. The stress values yielded by these two examples are compared between themselves and also with the results obtained by using commercial finite element packages to solve the same problems. Interesting differences between our results and those from the commercial packages are identified and reconciled.

There are aspects of finite element theory of interest to materials scientists which are not addressed in this book. Possibly the most important is the analysis of non-linear problems. There are two kinds of non-linearity – geometrical and material. Geometrical non-linearity arises when the distortions produced by the applied loads change the dimensions of the deforming body significantly. In the analysis of mechanical working processes such as extrusion, rolling and deep drawing this of course happens. It is not difficult, in principle, to cope with this situation by finite element analysis. If the loads are applied in small increments such that the distortion produced by each increment is not excessive, the deformation generated by one load increment can be calculated by the elastic analysis we develop here. Then the final dimensions when the load increment is complete become the starting dimensions for the next load increment. So by a judicious choice of the size of the load increments a useful solution can be achieved.

Material non-linearity arises in deformation processes where, as in metals, the flow stress varies with plastic strain or, in polymers, where it varies with strain rate. The analysis of such conditions requires the identification of a universal relation between a stress parameter and a strain parameter which applies during deformation under all stress states. This is a complex area. One approach for metallic plastic flow is to relate increments of octahedral shear stress to

3

increments of octahedral shear strain experimentally and use these data for finite element analysis using stepwise load increments.

The other topic that is omitted is the analysis of plates or shells by finite elements. Laterally loaded plates carry bending stresses and loaded shells have a mixture of membrane stresses and bending stresses. The finite element analysis of these cases present interesting issues that are beyond the scope of this introductory treatment.

We end this introduction by giving an example of the kind of problem that is amenable to finite element analysis. The problem has been chosen because an exact analytical solution exists. The finite element method gives approximate solutions. The identification of the degree of error in a particular solution is a complex matter. But if we choose a problem for which an exact solution exists and then work out a finite element solution the degree of error is precisely established by comparing the two solutions.

The problem is the classical one of the stress concentration around the hole in a sheet loaded in tension. The sheet is uniformly loaded at the two ends and, well away from the hole, the stress is uniform axial tension. Inglis produced the classical solution for this problem in a sheet of infinite width. This condition will be approached quite closely if the sheet width is more than ten times the hole diameter. The loaded sheet is illustrated in Fig. (1.2).

We set up a polar coordinate system (r, θ) with the zero θ axis running along the axis of the strip. The hole radius is r_0.

Inglis gives the circumferential stress round the hole to be

$$\sigma_{\theta\theta} = \frac{\sigma}{2}\left\{1 + \left(\frac{r_0}{r}\right)^2 - \left[1 + 3\left(\frac{r_0}{r}\right)^4\right]\cos(2\theta)\right\}$$

where σ is the axial stress remote from the hole.

We are interested in the axial tensile stress over the plane a–b through the centre of the hole,

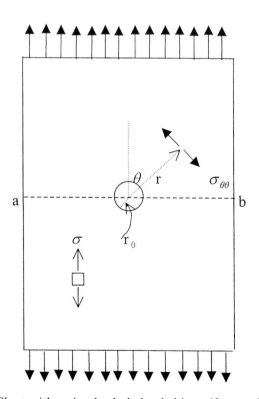

Fig. 1.2   Sheet with a circular hole loaded in uniform axial tension.

which is the value of $\sigma_{\theta\theta}$ at $\theta = \pi/2$. At this value of θ

$$\sigma_{\theta\theta} = \frac{\sigma}{2}\left[2 + \left(\frac{r_0}{r}\right)^2 + 3\left(\frac{r_0}{r}\right)^4\right] \tag{1.1}$$

At $r = r_0$, the surface of the hole, $\sigma_{\theta\theta}/\sigma = 3$, the well-known result.

The variation of $\sigma_{\theta\theta}/\sigma$ with r/r_0 from eqn (1.1) is graphed in Fig. 1.3 using square points.

A finite element solution to the same problem was obtained using the ABAQUS package. The chosen mesh of constant strain triangular elements is shown in Fig. 1.4. The mesh for a quarter of the plate is displayed. The width of the sheet is eleven times the hole diameter so the solution should be close to the Inglis values for an infinitely wide plate. An evident feature of the mesh is the high density of elements around the hole where the stress is varying most steeply. The individual elements near the hole cannot be resolved in the upper diagram of Fig. 1.4 so a magnified picture of this region is presented in the lower half of the figure to reveal the element pattern. The values of the axial stress ratio $\sigma_{\theta\theta}/\sigma$ in each element from the centre of the hole to the point b along a–b were extracted from the ABAQUS output and are graphed as diamond points in Fig. 1.3. Because the stress is constant over the element, we have plotted the stress at the point corresponding to the centre of the element.

The finite element values are seen to approach the Inglis solution closely but not exactly. This comparison emphasises that finite element solutions are approximate. The availability of the exact solution permits us to assert that this finite element is a worthwhile one. The stress concentration factor at the hole is 3.02. The maximum difference from the Inglis solution is a little over 5% at $r/r_0 = 1.5$.

We shall, in Chapter 8, carry out an analysis using an 8-node element. This makes it possible to use an interpolation function that gives a linear variation of stress over the element and, through this, yields solutions that have a smaller degree of error than those produced using the constant stress triangular element. The 8-node element will achieve this improvement even with a smaller number of elements.

We illustrate the capacity of the 8-node element to produce stress values of higher accuracy by obtaining a solution to the hole-in-the-plate problem using this element in the ABAQUS package. The element is designated CPS8R, indicating an 8-node element in plane stress using reduced integration, with four integration points, which we explain in detail later.

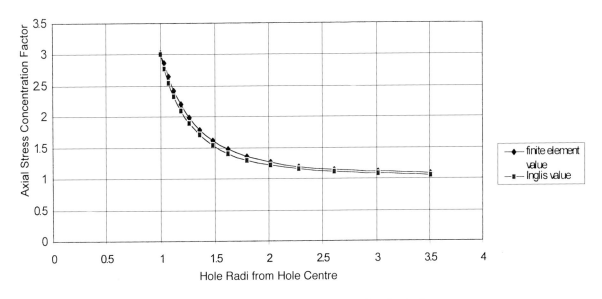

Fig. 1.3 Stress concentration at circular hole.

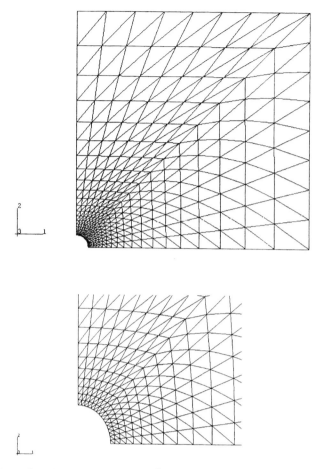

Fig. 1.4 Mesh of triangular elements shown for $\frac{1}{4}$ of the plate with a central hole. Lower diagram shows a magnified picture of the mesh close to the hole.

The element mesh is shown in Fig. 1.5 with a supplementary magnified view of the fine mesh round the hole. The quadrilateral mesh has about half the number of elements as the triangular mesh of Fig. 1.4.

The following table compares the ABAQUS and the exact Inglis values of the axial stress ratio $(\sigma_{\theta\theta}/\sigma)$ over the plane a–b (see Fig. 1.2) at selected values of the distance from the centre of the hole. The difference between the two never exceeds about 1% so the error in this solution using quadrilateral elements is about $1/5^{th}$ of that for the constant strain triangular element. This illustrates the superior potential of the 8-node element.

Values of the Axial Stress Concentration Factor $(\sigma_{\theta\theta}/\sigma)$ on the Plane through the Hole

Normalised distance from hole centre (r/r_0)	ABAQUS Stress Factor	Inglis Stress Factor
1.00	3.025	3.000
1.028	2.852	2.818
1.055	2.677	2.658
1.125	2.197	2.173
1.211	2.047	2.038
1.455	1.575	1.572
2.264	1.162	1.154
3.680	1.051	1.045
11.85	1.005	1.004
20.0	0.992	1.0012

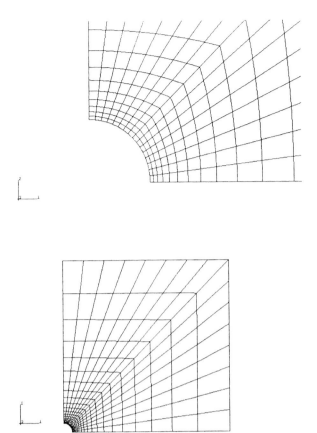

Fig. 1.5   The ABAQUS mesh of quadrilateral elements

CHAPTER 2   ABOUT MATRICES

2.1   WHAT IS A MATRIX?

The finite element method generates sets of simultaneous equations relating a large number of variables. The most orderly and compact way of presenting and of solving these equations is to use matrix representation. It is therefore essential to develop some feel for matrices. Here is a simple example. We preserve an element of nostalgia by working in Imperial units.

Consider the cost y_1 in week 1 of a basket of groceries comprising x_1 lb of cheese, x_2 lb of butter, x_3 lb of sugar and x_4 lb of tea. In week 1 the cost per pound of each item was:

cheese c_{11} butter c_{12} sugar c_{13} and tea c_{14}.

So the cost of the grocery basket in week 1 was :

$$y_1 = c_{11}x_1 + c_{12}x_2 + c_{13}x_3 + c_{14}x_4.$$

Assume that the same items were bought each week but the cost per pound changed. In week 2 the costs per pound were:

cheese c_{21} butter c_{22} sugar c_{23} and tea c_{24}.

The cost of the grocery basket in week 2 was:

$$y_2 = c_{21}x_1 + c_{22}x_2 + c_{23}x_3 + c_{24}x_4.$$

Correspondingly the costs in weeks 3 and 4 would be:

$$y_3 = c_{31}x_1 + c_{32}x_2 + c_{33}x_3 + c_{34}x_4$$

$$y_4 = c_{41}x_1 + c_{42}x_2 + c_{43}x_3 + c_{44}x_4.$$

These four simultaneous equations are displayed in matrix form as

$$\begin{Bmatrix} y_1 \\ y_2 \\ y_3 \\ y_4 \end{Bmatrix} = \begin{bmatrix} c_{11} & c_{12} & c_{13} & c_{14} \\ c_{21} & c_{22} & c_{23} & c_{24} \\ c_{31} & c_{32} & c_{33} & c_{34} \\ c_{41} & c_{42} & c_{43} & c_{44} \end{bmatrix} \begin{Bmatrix} x_1 \\ x_2 \\ x_3 \\ x_4 \end{Bmatrix} \tag{2.1}$$

The array of coefficients, or operators, in the square brackets is called a matrix. The order of a matrix is specified by the number of rows (r) and columns (c). So the order is $[r \times c]$ and the matrix above is $[4 \times 4]$. In this case $r = c$, which is called a square matrix, but r and c need not be equal as we shall see in many examples later. The original equations can be retrieved by multiplying out the matrix. For example, to retrieve the equation for the cost of the shopping basket in week 2, y_2, we multiply the coefficients in the row containing y_2 by the variables x_1 to x_4

according to the following scheme:

$$
\left\{
\begin{array}{c}
\bullet \\
y_2 \\
\bullet \\
\bullet
\end{array}
\right\}
=
\left[
\begin{array}{cccc}
\bullet & \bullet & \bullet & \bullet \\
c_{21} & c_{22} & c_{23} & c_{24} \\
\bullet & \bullet & \bullet & \bullet \\
\bullet & \bullet & \bullet & \bullet
\end{array}
\right]
\left\{
\begin{array}{c}
x_1 \\
x_2 \\
x_3 \\
x_4
\end{array}
\right\}
$$

so $y_2 = c_{21}x_1 + c_{22}x_2 + c_{23}x_3 + c_{24}x_4$.

2.2 MATRIX MULTIPLICATION

It is possible to multiply two matrices together only if the number of columns in the first matrix is equal to the number of rows in the second. So , for example, a matrix with 5 rows and 3 columns can be multiplied by a matrix with 3 rows and 2 columns . We shall see that the product is a $[5 \times 2]$ matrix. In general a matrix of order $[k \times j]$ multiplied by a matrix of order $[j \times i]$ will yield a matrix of order $[k \times i]$.

The scheme for multiplication is as follows:

$$
\left[
\begin{array}{ccc}
\circ & \circ & \circ \\
\circ & \circ & \circ \\
\circ & \circ & \circ \\
\circ & \circ & \circ \\
\circ & \circ & \circ
\end{array}
\right] X
\left[
\begin{array}{cc}
\circ & \circ \\
\circ & \circ \\
\circ & \circ
\end{array}
\right]
=
\left[
\begin{array}{cc}
\bullet & \otimes \\
\circ & \circ \\
\circ & \circ \\
\circ & \circ \\
\circ & \circ
\end{array}
\right]
\tag{2.2}
$$

The term that occupies $\bullet$ in the product matrix is obtained by summing the products of the terms connected by lines between the two multiplied matrices. Correspondingly the term in location $\otimes$ is obtained by summing the following products:

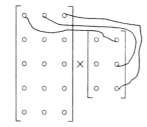

When the process is repeated for the remaining four rows of the left-hand matrix a $[5 \times 2]$ matrix is generated as the product. This is displayed in eqn (2.2).

The following example illustrates the multiplication process:

$$
\begin{bmatrix} 3 & 1 & 0 \\ 2 & 5 & 1 \\ 0 & 4 & 2 \\ 6 & 1 & 5 \\ 3 & 0 & 3 \end{bmatrix} \times \begin{bmatrix} 1 & 0 \\ 2 & 3 \\ 0 & 4 \end{bmatrix} = \begin{bmatrix} 5 & 3 \\ 12 & 19 \\ 8 & 20 \\ 8 & 23 \\ 3 & 12 \end{bmatrix}
$$

It is important to realise that if we reverse the order of two matrices that are multiplied we change the result. In particular if we reverse the matrices of the previous example:

$$
\begin{bmatrix} 1 & 0 \\ 2 & 3 \\ 0 & 4 \end{bmatrix} \times \begin{bmatrix} 3 & 1 & 0 \\ 2 & 5 & 1 \\ 0 & 4 & 2 \\ 6 & 1 & 5 \\ 3 & 0 & 3 \end{bmatrix}
$$

the multiplication process is not defined because the number of columns in the first matrix is not the same as the number of rows in the second matrix. If, by way of a further example, we take two square matrices, where multiplication is legal when the matrix order is reversed,

$$
\begin{bmatrix} 1 & 3 & 0 \\ 0 & 1 & 2 \\ 3 & 2 & 5 \end{bmatrix} \times \begin{bmatrix} 2 & 1 & 1 \\ 1 & 3 & 0 \\ 0 & 4 & 3 \end{bmatrix} = \begin{bmatrix} 5 & 10 & 1 \\ 1 & 11 & 6 \\ 8 & 29 & 18 \end{bmatrix} \quad \text{but}
$$

$$
\begin{bmatrix} 2 & 1 & 1 \\ 1 & 3 & 0 \\ 0 & 4 & 3 \end{bmatrix} \times \begin{bmatrix} 1 & 3 & 0 \\ 0 & 1 & 2 \\ 3 & 2 & 5 \end{bmatrix} = \begin{bmatrix} 5 & 9 & 7 \\ 1 & 6 & 6 \\ 9 & 10 & 23 \end{bmatrix}
$$

Evidently we get different products if the matrix order is reversed. In mathematical vocabulary matrix multiplication is not commutative.

2.3   TRANSPOSE OF A MATRIX

We will come across the transpose of a matrix which is achieved by interchanging the rows and columns. So the transpose of

$$
C = \begin{bmatrix} a & b & c \\ d & e & f \\ g & h & k \end{bmatrix} \quad \text{is} \quad C^T = \begin{bmatrix} a & d & g \\ b & e & h \\ c & f & k \end{bmatrix}.
$$

In finite element analysis the variables, for example forces or displacements , are usually vectors (quantities having both magnitude and direction). So if x is such a variable,

$$\begin{Bmatrix} x_1 \\ x_2 \\ x_3 \\ x_4 \end{Bmatrix}$$

would be called a column vector and is conventionally enclosed in curly brackets. The transpose of a column vector is a row vector, that is

$$\begin{Bmatrix} x_1 \\ x_2 \\ x_3 \\ x_4 \end{Bmatrix}^T = \{ x_1 \quad x_2 \quad x_3 \quad x_4 \}.$$

Column vectors are often presented as the transpose of row vectors just to save space on the printed page. Thus

$$\{ x_1 \quad x_2 \quad x_3 \quad x_4 \}^T = \begin{Bmatrix} x_1 \\ x_2 \\ x_3 \\ x_4 \end{Bmatrix}.$$

2.4   MATRIX INVERSION

The shopping basket matrix equation is of the form:

$$\{y\} = [C]\{x\}.$$

So given the quantities of each item purchased (the four values of x) and the matrix of costs per pound $[C]$ we could work out the four weekly values of y the cost of the food basket. If alternatively we know the cost of the food basket each week and wish to work out how much of each item was purchased (the values of x) then we would need a relation of the form:

$$\{x\} = [C]^{-1} \{y\}. \quad [C]^{-1} \text{ is called the inverse of matrix } [C].$$

The procedures for calculating the matrix inverse are tedious. We need to use two of these, which we now present.

2.4.1   FORMAL METHOD OF MATRIX INVERSION

The following classical procedure for matrix inversion is never used in computer programmes but we shall use it in a section of the analysis in Chapter 7 so it is helpful to explain it here. To invert the matrix

$$
\begin{bmatrix}
a_{11} & a_{12} & a_{13} \\
a_{21} & a_{22} & a_{23} \\
a_{31} & a_{32} & a_{33}
\end{bmatrix}
$$

we first set up the matrix of co-factors

$$
\begin{bmatrix}
A_{11} & A_{12} & A_{13} \\
A_{21} & A_{22} & A_{23} \\
A_{31} & A_{32} & A_{33}
\end{bmatrix}.
$$

The co-factor of a_{11} is A_{11} and is calculated by first working out the value of the determinant formed by deleting all the terms in the row and the column in which a_{11} lies. This determinant is

$$
\begin{bmatrix}
a_{22} & a_{23} \\
a_{32} & a_{33}
\end{bmatrix}
\qquad \text{and is called the minor of } a_{11}.
$$

A determinant is the value of an array of terms in a matrix obtained by multiplying the terms out in a prescribed way. The determinant of the above array is

$$
a_{22}\,a_{33} - a_{32}\,a_{23}.
$$

Finally the co-factor is obtained by multiplying the determinant by either $+1$ or -1 according to the position of the co-factor in the matrix according to the following scheme:

$$
\begin{bmatrix}
+1 & -1 & +1 \\
-1 & +1 & -1 \\
+1 & -1 & +1
\end{bmatrix}.
$$

The matrix inverse is then obtained by first transposing the co-factor matrix to

$$
\begin{bmatrix}
A_{11} & A_{21} & A_{31} \\
A_{12} & A_{22} & A_{32} \\
A_{13} & A_{23} & A_{33}
\end{bmatrix}
$$

and then dividing this by the determinant of the original matrix, which is

$$
D = a_{11}(a_{22}a_{33} - a_{32}a_{23}) - a_{12}(a_{21}a_{33} - a_{31}a_{23}) + a_{13}(a_{21}a_{32} - a_{31}a_{22}).
$$

Let us illustrate this procedure by working through a simple example. Find the inverse of the matrix

$$
\begin{bmatrix}
1 & 7 & 4 \\
7 & 2 & 5 \\
4 & 5 & 3
\end{bmatrix}
$$

13

The co-factor of the first term in the first row is $+1(2 \times 3 - 5 \times 5) = -19$ and the co-factor of the second term in that row is $-1(7 \times 3 - 5 \times 4) = -1$. The complete set of co-factors is

$$
\begin{bmatrix}
-19 & -1 & 27 \\
-1 & -13 & 23 \\
27 & 23 & -47
\end{bmatrix}
\quad \text{and the transpose is} \quad
\begin{bmatrix}
-19 & -1 & 27 \\
-1 & -13 & 23 \\
27 & 23 & -47
\end{bmatrix}.
$$

We divide the transpose by the determinant D of the original matrix to obtain the inverse

$$
D = 1(3 \times 2 - 5 \times 5) - 7(7 \times 3 - 4 \times 5) + 4(7 \times 5 - 4 \times 2) = 82.
$$

To divide a matrix by a constant we divide each term in the matrix by that constant so the matrix inverse is

$$
\begin{bmatrix}
-19/82 & -1/82 & 27/82 \\
-1/82 & -13/82 & 23/82 \\
27/82 & 23/82 & -47/82
\end{bmatrix}.
$$

2.4.2 GAUSSIAN ELIMINATION

A process that is appropriate for computer processing to invert matrix equations is Gaussian elimination. It works on the following principle:

Consider the following matrix equation

$$
\begin{bmatrix}
c_{11} & c_{12} & c_{13} & c_{14} \\
c_{21} & c_{22} & c_{23} & c_{24} \\
c_{31} & c_{32} & c_{33} & c_{34} \\
c_{41} & c_{42} & c_{43} & c_{44}
\end{bmatrix}
\begin{Bmatrix}
x_1 \\
x_2 \\
x_3 \\
x_4
\end{Bmatrix}
=
\begin{Bmatrix}
y_1 \\
y_2 \\
y_3 \\
y_4
\end{Bmatrix}
\tag{2.2}
$$

Assume that we know the four y values and the terms in the $[4 \times 4]$ matrix and wish to work out the four values of the variable x. To this end we modify the matrix (2.2) and transform it into the following form

$$
\begin{bmatrix}
a & b & c & d \\
0 & e & f & g \\
0 & 0 & h & k \\
0 & 0 & 0 & l
\end{bmatrix}
\begin{Bmatrix}
x_1 \\
x_2 \\
x_3 \\
x_4
\end{Bmatrix}
=
\begin{Bmatrix}
my_1 \\
ny_2 \\
py_3 \\
ry_4
\end{Bmatrix}.
$$

If we multiply out the fourth equation, that is the fourth row,

$$
(0.x_1) + (0.x_2) + (0.x_3) + (l.x_4) = ry_4 \quad \text{so } x_4 = ry_4/l.
$$

now multiply out the third equation (row 3) ignoring the zero terms

$$
(h.x_3) + (k.x_4) = py_3 \quad \text{but we know } x_4 \text{ so } x_3 \text{ is determined.}
$$

14

Similarly from the second equation (row 2)

$$(e.x_2) + (f.x_3) + (g.x_4) = ny_2 \quad \text{we determine } x_2$$

and from the first equation (row 1)

$$(a.x_1) + (b.x_2) + (c.x_3) + (d.x_4) = my_1 \quad \text{we determine } x_1.$$

The principle that is used to modify the matrix to generate zeros below what is called the leading diagonal (*aehl*) is the following. The balance of any equation is not upset if both sides are multiplied by the same factor. Similarly the balance is preserved if the same quantity is either added to or subtracted from both sides of the equation. So if we take the third equation that is row 3 of the original matrix (2.2) and multiply it by $\frac{c_{41}}{c_{31}}$ we get

$$\frac{c_{41}}{c_{31}} c_{31} x_1 + \frac{c_{41}}{c_{31}} c_{32} x_2 + \frac{c_{41}}{c_{31}} c_{33} x_3 + \frac{c_{41}}{c_{31}} c_{34} x_4 = \frac{c_{41}}{c_{31}} y_3$$

The first term in this equation is $c_{41}.x_1$ so if we modify equation 4, that is row 4 of (2.2), by subtracting the equation from it, the first term in equation 4 becomes zero and we have generated the first desired zero below the leading diagonal. By choosing appropriate pairs of equations, that is appropriate pairs of rows, and multiplying one of them by an appropriate factor and adding it to or subtracting it from the other we can generate additional zeros and in this way achieve a matrix in which all the terms below the leading diagonal are zero. Let us illustrate the method by working out a problem using the shopping basket that we introduced in section (2.1). Let us assume that we know the cost/lb of each of the four items in the shopping basket and we also know the bill in each week for the basket of food. Our task is to work out what weight of each item was bought, recalling that this was the same each week. The information we start with is as follows

COST/LB (£).

	Cheese (c)	Butter (b)	Ham (h)	Tomatoes (t)
Week1	2.76	1.17	3.10	0.34
Week2	2.62	1.26	3.32	0.42
Week3	2.96	1.09	3.70	0.56
Week4	2.53	1.33	3.52	0.37

COST OF FOOD BASKET (£)

Week 1	9.825
Week 2	10.20
Week 3	11.075
Week 4	10.205

Putting this information in the matrix equation (2.1)

$$\begin{bmatrix} 2.76 & 1.17 & 3.10 & 0.34 \\ 2.62 & 1.26 & 3.32 & 0.42 \\ 2.96 & 1.09 & 3.70 & 0.56 \\ 2.53 & 1.33 & 3.52 & 0.37 \end{bmatrix} \begin{Bmatrix} c \\ b \\ h \\ t \end{Bmatrix} = \begin{Bmatrix} 9.825 \\ 10.20 \\ 11.075 \\ 10.205 \end{Bmatrix}$$

In order to solve this matrix equation to obtain the values of the unknowns c, b, h and t we modify the matrix to produce zeros in all the terms below the leading diagonal.

15

Multiply row 3 by $\frac{2.53}{2.96}$ and subtract the modified row 3 from row 4. Row 3 becomes

$$[2.53 \quad 0.93165 \quad 3.1625 \quad 0.4786]\{\} = \{9.46613\}$$

and after subtracting this from row 4 the new row 4 becomes

$$[0 \quad 0.39834 \quad 0.3575 \quad -0.20865]\{\} = \{0.73886\}$$

Now multiply row 2 by $\frac{2.96}{2.62}$ it becomes

$$[2.96 \quad 1.4325 \quad 3.7508 \quad 0.4745]\{\} = \{11.52366\}$$

Subtract this from the original row 3 which then becomes

$$[0 \quad -.33351 \quad -0.05084 \quad 0.08549]\{\} = \{-0.44866\}$$

At this stage the matrix equation has been modified to

$$\begin{bmatrix} 2.76 & 1.17 & 3.10 & 0.34 \\ 2.62 & 1.26 & 3.32 & 0.42 \\ 0 & -0.33351 & -0.05084 & 0.08549 \\ 0 & 0.39834 & 0.35750 & -0.10865 \end{bmatrix} \begin{Bmatrix} c \\ b \\ h \\ t \end{Bmatrix} = \begin{Bmatrix} 9.825 \\ 10.200 \\ -0.44866 \\ 0.73886 \end{Bmatrix}$$

We now have two zeros below the leading diagonal. Now multiply row 3 by $\frac{0.39834}{0.33351}$ and row 3 becomes

$$[0 \quad -0.39834 \quad -0.06072 \quad 0.10211]\{\} = \{-0.53588\}$$

Add this to row 4, and row 4 becomes

$$[0 \quad 0 \quad 0.29677 \quad -0.00654]\{\} = \{0.20299\}$$

Now multiply row 1 by $\frac{2.62}{2.76}$; row 1 becomes

$$[2.62 \quad 1.11065 \quad 2.94275 \quad 0.32275]\{\} = \{9.32663\}$$

Subtract this from row 2; row 2 becomes

$$[0 \quad 0.14935 \quad 0.37225 \quad 0.09725]\{\} = \{0.87337\}$$

The matrix equation is now

$$\begin{bmatrix} 2.76 & 1.17 & 3.10 & 0.34 \\ 0 & 0.14935 & 0.37724 & 0.09725 \\ 0 & -0.33351 & -0.05084 & 0.08549 \\ 0 & 0 & 0.29677 & -0.00654 \end{bmatrix} \begin{Bmatrix} c \\ b \\ h \\ t \end{Bmatrix} = \begin{Bmatrix} 9.825 \\ 0.87337 \\ -0.44866 \\ 0.20299 \end{Bmatrix}$$

We now have four zeros below the main diagonal. We need to create two more to achieve our

16

objective. Multiply row 2 by $\frac{0.33351}{0.14935}$, row 2 becomes

$$[0 \quad 0.33351 \quad 0.8240 \quad 0.21717]\{\} = \{1.95030\}$$

Add this to row 3 and row 3 becomes

$$[0 \quad 0 \quad 0.79156 \quad 0.30266]\{\} = \{1.50164\}$$

multiply row 3 by $\frac{0.29677}{0.79156}$, row 3 becomes

$$[0 \quad 0 \quad 0.29677 \quad 0.11347]\{\} = \{0.56299\}$$

Subtract this from row 4 and row 4 becomes

$$[0 \quad 0 \quad 0 \quad -0.12001]\{\} = \{-0.36000\}$$

We now have the matrix in its final form:

$$\begin{bmatrix} 2.76 & 1.17 & 3.10 & 0.34 \\ 0 & 0.14935 & 0.37724 & 0.09725 \\ 0 & 0 & 0.79156 & 0.30266 \\ 0 & 0 & 0 & -0.12001 \end{bmatrix} \begin{Bmatrix} c \\ b \\ h \\ t \end{Bmatrix} = \begin{Bmatrix} 9.825 \\ 0.87337 \\ 1.50164 \\ -0.36000 \end{Bmatrix}$$

The value of t, the weight of tomatoes bought, can now be derived by multiplying out the fourth row

$$0.c + 0.b + 0.h - 0.12001.t = -0.36000 \quad \text{so } t = 2.9998 \text{ lb.}$$

Multiply out the third row, omitting the zero terms this time

$$0.79156.h + 0.30266.t = 1.50164$$

Substituting for t gives

$$h = 0.75006 \text{ lb.}$$

Multiply out the second row omitting the zero term

$$0.14935.b + 0.37724.h + 0.09725.t = 0.87337$$

Substituting for h and t yields

$$b = 1.9999 \text{ lb.}$$

Finally multiply out the top equation in row 1

$$2.76.c + 1.17.b + 3.10.h + 0.34.t = 9.825$$

Substituting for b, h and t yields

$$c = 1.4999 \text{ lb.}$$

These weights are very close to the exact values of

$$t = 3 \text{ lb.} \quad h = 0.75 \text{ lb.} \quad b = 2 \text{ lb.} \quad \text{and} \quad c = 1.5 \text{ lb.}$$

The calculated values would converge on the exact values as we increased the number of significant figures in the calculations.

CHAPTER 3   THE STIFFNESS MATRIX

The dominant feature of the finite element analytical process is the determination of the Stiffness Matrix. In the chapters that follow, we shall be much engaged in deriving this matrix for a number of different conditions. Before doing that it is helpful to lay out some general ideas about the nature of this important matrix and the significance of the terms within it. It is not necessary at this stage to know how the matrix is derived; that will come later and will involve much detailed analysis. Whilst we are involved in this analysis, the following general discussion of the Stiffness Matrix will allow us to place the details in a broader context. We begin, curiously, not by addressing finite elements but structural engineering.

3.1   PIN-JOINTED STRUCTURES

A structure formed from rods of uniform cross-section pinned together is a simple natural finite element system. Each rod is an element and the points at which the rods join are nodes. The fact that the nodes are pinned means that only tension or compression forces can be transmitted to the rods so the stresses in them are either uniform tension or compression. The fact that the pins cannot transmit moments means that there are no bending stress in the rods. An example of such a structure is;

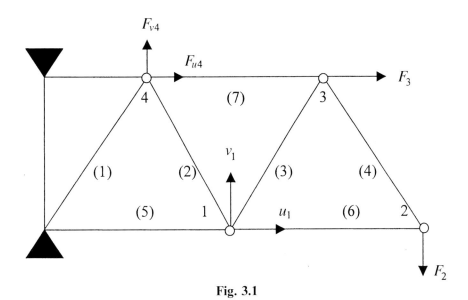

Fig. 3.1

The nodes are numbered. The element numbers are in parentheses. The two nodes at which the structure is supported and at which the displacements are constrained to be zero have been excluded from the numbering sequence. We are interested in the displacement of this elastic structure produced by applied loads. Two such loads are F_2 at node 2 and F_3 at node 3. The nodal displacements are specified by the horizontal and vertical components, for example u_1 and v_1 at node 1. Each displacement component is called a degree of freedom. Since there are four nodes each with two degrees of freedom the structure has eight degrees of freedom. The nodal forces are also expressed in terms of the horizontal and vertical components, for example F_{u4} and F_{v4} at node 4.

Structural engineers have analysed problems like this in the past using the Flexibility Method. This is based on the following assumptions:

(i) Any load component will produce displacements in all degrees of freedom of the structure.
(ii) Each displacement varies linearly with the force producing it
(iii) The deflections generated by a number of loads is the sum of the deflections that would be produced by each load acting singly.

We now apply these principles to the structure in Fig. 3.1. It is helpful to number the degrees of freedom in sequence according to the following scheme

Displacement component	u_1	v_1	u_2	v_2	u_3	v_3	u_4	v_4
Degree of freedom	1	2	3	4	5	6	7	8

Consider first the nodal displacement u_1 at node 1. We assume that it comprises the sum of the displacements generated by all the eight force components which act in the direction of and at the location of the eight degrees of freedom. So we can express

$$u_1 = u_{11} + u_{12} + u_{13} + u_{14} + u_{15} + u_{16} + u_{17} + u_{18} \qquad (3.1)$$

u_{11} is the contribution to u_1 from the force corresponding to degree of freedom 1 that is F_{u1}. u_{12} is the contribution to u_1 from the force corresponding to degree of freedom 2 that is F_{v1}. u_{16} is the contribution to u_1 from the force corresponding to degree of freedom 6 that is F_{v3} and so on.

There is a linear relation between the forces and the displacement components so

$$u_{11} = c_{11} F_{u1} \qquad u_{12} = c_{12} F_{v1} \qquad u_{16} = c_{16} F_{v3} \qquad (3.2)$$

Now $c_{11} = u_{11}/F_{u1}$ that is displacement/force which is a flexibility so c_{11} is called a Flexibility Coefficient. c_{11} is the displacement produced at u_1 by a unit force F_{u1} and c_{16} is the displacement produced at u_1 by a unit force F_{v3}.

We can substitute relations like those in eqns (3.2) for all the displacement components in eqn 3.1 and then

$$u_1 = c_{11}F_{u1} + c_{12}F_{v1} + c_{13}F_{u2} + c_{14}F_{v2} + c_{15}F_{u3} + c_{16}F_{v3} + c_{17}F_{u4} + c_{18}F_{v4} \qquad (3.3)$$

There will be relations like eqn 3.1 for all the eight displacements for example

$$v_1 = u_{21} + u_{22} + u_{23} + u_{24} + u_{25} + u_{26} + u_{27} + u_{28} \qquad (3.4)$$

where for example u_{25} is the displacement at degree of freedom 2, that is v_1, produced by a force at degree of freedom 5 (that is F_{u3}). Incorporating the appropriate flexibility coefficients in eqn 3.4 gives

$$v_1 = c_{21}F_{u1} + c_{22}F_{v1} + c_{23}F_{u2} + c_{24}F_{v2} + c_{25}F_{u3} + c_{26}F_{v3} + c_{27}F_{u4} + c_{28}F_{v4} \qquad (3.5)$$

Similarly

$$u_2 = c_{31}F_{u1} + c_{32}F_{v1} + c_{33}F_{u2} + c_{34}F_{v2} + c_{35}F_{u3} + c_{36}F_{v3} + c_{37}F_{u4} + c_{38}F_{v4} \qquad (3.6)$$

There are five more equations like these for the complete description of the displacements of the structure. Each equation has 8 flexibility coefficients so there are $8 \times 8 = 64$ coefficients in all. In general a structure with n degrees of freedom will have n^2 flexibility coefficients.

The eight equations are expressed in a compact fashion using a matrix format:

$$\begin{Bmatrix} u_1 \\ v_1 \\ u_2 \\ v_2 \\ u_3 \\ v_3 \\ u_4 \\ v_4 \end{Bmatrix} = \begin{bmatrix} c_{11} & c_{12} & c_{13} & c_{14} & c_{15} & c_{16} & c_{17} & c_{18} \\ c_{21} & c_{22} & c_{23} & c_{24} & c_{25} & c_{26} & c_{27} & c_{28} \\ c_{31} & c_{32} & c_{33} & c_{34} & c_{35} & c_{36} & c_{37} & c_{38} \\ c_{41} & c_{42} & c_{43} & c_{44} & c_{45} & c_{46} & c_{47} & c_{48} \\ c_{51} & c_{52} & c_{53} & c_{54} & c_{55} & c_{56} & c_{57} & c_{58} \\ c_{61} & c_{62} & c_{63} & c_{64} & c_{65} & c_{66} & c_{67} & c_{68} \\ c_{71} & c_{72} & c_{73} & c_{74} & c_{75} & c_{76} & c_{77} & c_{78} \\ c_{81} & c_{82} & c_{83} & c_{84} & c_{85} & c_{86} & c_{87} & c_{88} \end{bmatrix} \begin{Bmatrix} F_{u1} \\ F_{v1} \\ F_{u2} \\ F_{v2} \\ F_{u3} \\ F_{v3} \\ F_{u4} \\ F_{v4} \end{Bmatrix} \qquad (3.7)$$

$$\downarrow \qquad\qquad \downarrow \qquad\qquad \downarrow$$

NODAL   FLEXIBILITY MATRIX   NODAL
DISPLACEMENT   FORCE
VECTOR   VECTOR
$\{u\} =$ $[C]$ $\{F\}$

The matrix eqn 3.7 has three parts which are named in 3.7; the nodal displacement vector which is the array of all the displacement degrees of freedom, the flexibility matrix with its 64 flexibility coefficients and the nodal force vector which is the array of the eight force components. The equation has the structure $\{u\} = [C] \{F\}$.

A proof by Maxwell known as the Reciprocal Theorem asserts that in any structure the displacement produced at a point say 2 by a force at another point 4 is the same as the displacement at 4 produced by the same force at 2. The effect of this principle on the flexibility matrix in eqn (3.7) is to cause pairs of coefficients which have the same numerical suffices but in inverse order to be equal, for example $c_{42} = c_{24}$. So coefficients that are in mirror image positions in relation to the main diagonal of the matrix are equal. This reduces the number of coefficients needed to define the matrix from 64 to 36. A matrix that has this structure is called symmetric.

If we could calculate all the values of c_{ij} in (3.7) and we knew the nodal applied forces then we could solve (3.7) for the nodal displacements. This is a feasible but tedious process for frameworks like that in Fig. 3.1. However it turns out not to be a way forward for the finite element meshes that are our concern. Rather an inverse relationship is used which comprises the Stiffness Method.

3.2 THE STIFFNESS METHOD

The Stiffness Method is based on a relation that is the inverse of that for the Flexibility Method. We have seen that the Flexibility Method equation has the structure

$$\{u\} = [C] \{F\}.$$

In contrast, the Stiffness Method yields an equation of the form

$$[K] \{u\} = \{F\}. \qquad (3.8)$$

[K] is the Stiffness Matrix and, like the Flexibility matrix, for a structure with 8 degrees of freedom has 64 terms. So if we expand eqn (3.8) fully we get

$$
\begin{bmatrix}
k_{11} & k_{12} & k_{13} & k_{14} & k_{15} & k_{16} & k_{17} & k_{18} \\
k_{21} & k_{22} & k_{23} & k_{24} & k_{25} & k_{26} & k_{27} & k_{28} \\
k_{31} & k_{32} & k_{33} & k_{34} & k_{35} & k_{36} & k_{37} & k_{38} \\
k_{41} & k_{42} & k_{43} & k_{44} & k_{45} & k_{46} & k_{47} & k_{48} \\
k_{51} & k_{52} & k_{53} & k_{54} & k_{55} & k_{56} & k_{57} & k_{58} \\
k_{61} & k_{62} & k_{63} & k_{64} & k_{65} & k_{66} & k_{67} & k_{68} \\
k_{71} & k_{72} & k_{73} & k_{74} & k_{75} & k_{76} & k_{77} & k_{78} \\
k_{81} & k_{82} & k_{83} & k_{84} & k_{85} & k_{86} & k_{87} & k_{88}
\end{bmatrix}
\begin{Bmatrix} u_1 \\ v_1 \\ u_2 \\ v_2 \\ u_3 \\ v_3 \\ u_4 \\ v_4 \end{Bmatrix}
=
\begin{Bmatrix} F_{u1} \\ F_{v1} \\ F_{u2} \\ F_{v2} \\ F_{u3} \\ F_{v3} \\ F_{u4} \\ F_{v4} \end{Bmatrix}
\tag{3.9}
$$

We can develop an understanding of the significance of this matrix equation if we multiply out one of the constituent equations, for example the third row

$$k_{31}u_1 + k_{32}v_1 + k_{33}u_2 + k_{34}v_2 + k_{35}u_3 + k_{36}v_3 + k_{37}u_4 + k_{38}v_4 = F_{u2}. \tag{3.10}$$

The force on the right-hand side of eqn (3.10) is that at node 2 in the direction of the displacement u_2. All the terms on the left-hand side must also be forces. They are the contributions to F_{u2} arising from each of the eight displacements u_1 to v_4. So for example $k_{32} v_1$ is the force transmitted to node 2 in the direction of u_2 by a displacement v_1 at node 1. Similarly $k_{38} v_4$ is the force transmitted to node 2 in the direction of u_2 by a displacement v_4 at node 4.

It is evident therefore that an individual eqn in (3.9) such as (3.10) represents the total force transmitted to a particular node in a particular direction by displacements at all the nodes in the structure. This has the inverse sense of an equation from the matrix eqn (3.7) for the Flexibility Method which represents the total displacement at a particular node in a particular direction arising from forces at all the nodes in the structure.

There is however an extremely important difference between the flexibility matrix of (3.7) and the stiffness matrix of (3.9). The flexibility matrix admits that all forces in the structure can contribute to a particular displacement. In contrast we shall see that in the finite element method the derivation of the terms in the stiffness matrix makes the assumption that the only nodes that can transmit forces to a particular node are those that are nearest-neighbour to the node in question. This arises because in the analysis of the force transmission it is assumed that when a particular node is displaced all other nodes in the structure are fixed and therefore their displacement is zero. To better understand this principle consider the structure in Fig. 3.2.

It comprises an arrangement of elastic rods connected at nodes that are numbered 1 to 8. Displace node 3 by u_3 whilst all nodes except 3 are fixed. Evidently in this circumstance forces are transmitted to nodes 1, 2 and 4 because rods 1–3, 2–3 and 3–4 all change in length, but no forces are transmitted to nodes 5, 6, 7 and 8 because none of the rods connected to any of these nodes changes length. So only nodes that are nearest-neighbour to node 3 experience a force. By the same principle, the only nodes that can transmit a force to node 3 are nodes 1, 2 and 4. For example if node 4 is displaced with all other nodes fixed a force is transmitted to node 3 because 4–3 changes length, but if node 6 is displaced with nodes 1, 2, 3, 4, 5, 7 and 8 fixed no force is transmitted to node 3 because none of the rods connected to it changes length. The effect of this condition is to give rise to many zeros in the stiffness matrix. We analyse a structure in the next section to illustrate how the zeros are distributed in the stiffness matrix in two particular cases.

It is interesting to note that the structural engineers now favour the stiffness method for structural analysis because it lends itself more readily to computer analysis than does the

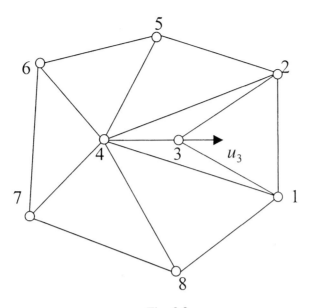

Fig. 3.2

flexibility method. The change has come about through the dramatic increase of computer power over recent decades. It is this that makes it practicable to contemplate the use of the stiffness method.

3.3 A COMPUTATIONAL POINT

The position of the zeros in the stiffness matrix has a significant effect on the computer time needed to perform a finite element analysis. The position of the zeros is affected by the sequence of numbering the nodes. We can illustrate this by a simple example. Consider a simple pin-jointed truss with 8 active nodes:

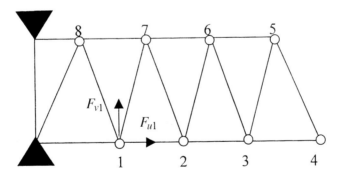

Fig. 3.3

The nodes are numbered in the sequence indicated. Now consider which nodal displacements will contribute to the forces at node 1, F_{u1} and F_{v1}. From our earlier considerations only displacements of the nearest-neighbour nodes 2, 7 and 8 can do so. The displacements at these nodes are u_2 v_2 u_7 v_7 u_8 and v_8 so the stiffness coefficients multiplying these displacements in the first two rows of the stiffness matrix, which are the equations for F_{u1} and F_{v1}, will be non-zero; all the other coefficients in those two rows will be zero. It is instructive to set up the stiffness matrix, see (3.11), and indicate the zero and non-zero terms for the structure in Fig 3.3. There are eight active nodes in the structure so there are 16 degrees of freedom and therefore the stiffness matrix is 16×16. In presenting the matrix equation we have added the appropriate nodal displacement component at the head of each column. This is a reminder that in the equations for the force components, the stress coefficient in a particular column is multiplied by the displacement component we have indicated at the head of it.

$$
\begin{array}{c}
\begin{array}{cccccccccccccccc} u_1 & v_1 & u_2 & v_2 & u_3 & v_3 & u_4 & v_4 & u_5 & v_5 & u_6 & v_6 & u_7 & v_7 & u_8 & v_8 \end{array} \\
\begin{bmatrix}
x & x & x & x & 0 & 0 & 0 & 0 & 0 & 0 & 0 & 0 & x & x & x & x \\
x & x & x & x & 0 & 0 & 0 & 0 & 0 & 0 & 0 & 0 & x & x & x & x \\
x & x & x & x & x & x & 0 & 0 & 0 & 0 & x & x & x & x & 0 & 0 \\
x & x & x & x & x & x & 0 & 0 & 0 & 0 & x & x & x & x & 0 & 0 \\
0 & 0 & x & x & x & x & x & x & x & x & x & x & 0 & 0 & 0 & 0 \\
0 & 0 & x & x & x & x & x & x & x & x & x & x & 0 & 0 & 0 & 0 \\
0 & 0 & 0 & 0 & x & x & x & x & x & x & 0 & 0 & 0 & 0 & 0 & 0 \\
0 & 0 & 0 & 0 & x & x & x & x & x & x & 0 & 0 & 0 & 0 & 0 & 0 \\
0 & 0 & 0 & 0 & x & x & x & x & x & x & x & x & 0 & 0 & 0 & 0 \\
0 & 0 & 0 & 0 & x & x & x & x & x & x & x & x & 0 & 0 & 0 & 0 \\
0 & 0 & x & x & x & x & 0 & 0 & x & x & x & x & x & x & 0 & 0 \\
0 & 0 & x & x & x & x & 0 & 0 & x & x & x & x & x & x & 0 & 0 \\
x & x & x & x & 0 & 0 & 0 & 0 & 0 & 0 & x & x & x & x & x & x \\
x & x & x & x & 0 & 0 & 0 & 0 & 0 & 0 & x & x & x & x & x & x \\
x & x & 0 & 0 & 0 & 0 & 0 & 0 & 0 & 0 & 0 & 0 & x & x & x & x \\
x & x & 0 & 0 & 0 & 0 & 0 & 0 & 0 & 0 & 0 & 0 & x & x & x & x
\end{bmatrix}
\begin{Bmatrix} u_1 \\ v_1 \\ u_2 \\ v_2 \\ u_3 \\ v_3 \\ u_4 \\ v_4 \\ u_5 \\ v_5 \\ u_6 \\ v_6 \\ u_7 \\ v_7 \\ u_8 \\ v_8 \end{Bmatrix}
=
\begin{Bmatrix} F_{u1} \\ F_{v1} \\ F_{u2} \\ F_{v2} \\ F_{u3} \\ F_{v3} \\ F_{u4} \\ F_{v4} \\ F_{u5} \\ F_{v5} \\ F_{u6} \\ F_{v6} \\ F_{u7} \\ F_{v7} \\ F_{u8} \\ F_{v8} \end{Bmatrix}
\end{array}
\qquad (3.11)
$$

If a stiffness coefficient is non-zero we shall indicate this by placing an 'x' at the appropriate location in the matrix; if the coefficient is zero we shall place a '0'. The first two rows of the stiffness matrix are the equations for the forces at node 1, F_{u1} and F_{v1}. We have seen that only displacements at nodes 2, 7 and 8, and of course at node 1 itself, can transmit a force to node 1. So the stiffness coefficients that multiply nodal displacements at nodes 1, 2, 7 and 8 are non-zero. These displacements are u_1, v_1, u_2, v_2, u_7, v_7, u_8, and v_8 so we have put an 'x' in rows 1 and 2 and in the columns headed by these displacements to identify this group of non-zero stiffness coefficients. All the other coefficients in rows 1 and 2 are zero and we have so indicated with '0' in all the remaining positions in the two rows of the matrix.

Rows 3 and 4 of the matrix are the equations that give the forces on node 2, which are F_{u2} and F_{v2}. The nodal displacements that can transmit a force to node 2 are those at the nearest-neighbour nodes 1, 2, 3, 6 and 7 so the non-zero stiffness coefficients are in the columns headed u_1 v_1 u_2 v_2 u_3 v_3 u_6 v_6 u_7 and v_7 and in rows 3 and 4. These are indicated by an 'x', the remaining coefficients in these two rows are '0'. We proceed in this manner for the other six nodes in the structure, that is the remaining 12 rows of the matrix and build up the array of zero and non-zero coefficients in the matrix displayed in eqn (3.11).

It is evident that the non-zero terms are distributed in the matrix in a broad cross pattern. The computer time needed to set up and invert a matrix turns out to depend on what is called the half band-width. This is the distance of the remotest non-zero term from the main diagonal. In the matrix above it is fifteen in rows 1 and 16.

Now let us establish the band-width for the matrix for the same structure as Fig. 3.3 but with the nodes numbered in the different sequence shown in Fig. 3.4.

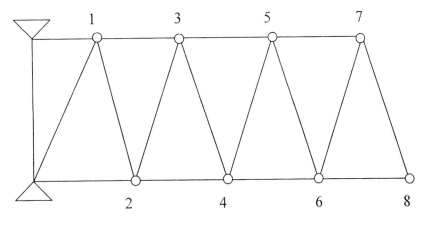

Fig. 3.4

We use the same method as we used to establish the non-zero terms in matrix equation (3.11), for example that only nodes 1, 2, 3 and 4 can transmit forces to node 2. In this way, we get the pattern of zero and non-zero terms displayed in the following matrix eqn (3.12).

With this node numbering sequence the non-zero terms are seen to occupy a single band around the leading diagonal. The half band-width is five so the computation time needed to set up and invert this matrix will be significantly less than for that of eqn (3.11). To achieve this desirable minimum band-width the node numbering sequence should be chosen so that nearest-neighbour nodes have the closest spread of numbers. Commercial finite element packages have regard to this principle.

$$
\begin{bmatrix}
x & x & x & x & x & x & 0 & 0 & 0 & 0 & 0 & 0 & 0 & 0 & 0 & 0 \\
x & x & x & x & x & x & 0 & 0 & 0 & 0 & 0 & 0 & 0 & 0 & 0 & 0 \\
x & x & x & x & x & x & x & x & 0 & 0 & 0 & 0 & 0 & 0 & 0 & 0 \\
x & x & x & x & x & x & x & x & 0 & 0 & 0 & 0 & 0 & 0 & 0 & 0 \\
x & x & x & x & x & x & x & x & x & x & 0 & 0 & 0 & 0 & 0 & 0 \\
x & x & x & x & x & x & x & x & x & x & 0 & 0 & 0 & 0 & 0 & 0 \\
0 & 0 & x & x & x & x & x & x & x & x & x & x & 0 & 0 & 0 & 0 \\
0 & 0 & x & x & x & x & x & x & x & x & x & x & 0 & 0 & 0 & 0 \\
0 & 0 & 0 & 0 & x & x & x & x & x & x & x & x & x & x & 0 & 0 \\
0 & 0 & 0 & 0 & x & x & x & x & x & x & x & x & x & x & 0 & 0 \\
0 & 0 & 0 & 0 & 0 & 0 & x & x & x & x & x & x & x & x & x & x \\
0 & 0 & 0 & 0 & 0 & 0 & x & x & x & x & x & x & x & x & x & x \\
0 & 0 & 0 & 0 & 0 & 0 & 0 & 0 & x & x & x & x & x & x & x & x \\
0 & 0 & 0 & 0 & 0 & 0 & 0 & 0 & x & x & x & x & x & x & x & x \\
0 & 0 & 0 & 0 & 0 & 0 & 0 & 0 & 0 & 0 & x & x & x & x & x & x \\
0 & 0 & 0 & 0 & 0 & 0 & 0 & 0 & 0 & 0 & x & x & x & x & x & x
\end{bmatrix}
\begin{Bmatrix}
u_1 \\ v_1 \\ u_2 \\ v_2 \\ u_3 \\ v_3 \\ u_4 \\ v_4 \\ u_5 \\ v_5 \\ u_6 \\ v_6 \\ u_7 \\ v_7 \\ u_8 \\ v_8
\end{Bmatrix}
=
\begin{Bmatrix}
F_{u1} \\ F_{v1} \\ F_{u2} \\ F_{v2} \\ F_{u3} \\ F_{v3} \\ F_{u4} \\ F_{v4} \\ F_{u5} \\ F_{v5} \\ F_{u6} \\ F_{v6} \\ F_{u7} \\ F_{v7} \\ F_{u8} \\ F_{v8}
\end{Bmatrix}
\qquad (3.12)
$$

CHAPTER 4   ONE-DIMENSIONAL FINITE ELEMENT ANALYSIS

A coiled spring is a simple finite element. There is a simple relation between the tension in the spring and its extension. It turns out that for such a simple element the displacements produced when an assembly of such springs is loaded can be analysed using the following three principles:

1. Connected ends of adjacent springs have equal displacements. This is called the condition of compatibility.
2. The resultant force at each node is zero. This is the condition of static equilibrium.
3. The extension of the spring is proportional to the tension in it. This is called the constitutive relation.

These principles suffice for the solution of many problems in the stress analysis of loaded elastic bodies. However we shall see that when we come to analyse the stresses and strains in elastic continua using the finite element method, these principles do not suffice and we shall have to introduce an additional principle based on elastic and potential energy. This is needed specifically to relate the forces on the element to the stresses within it.

The analysis of an assembly of springs does not need this additional complication so we can use such an analysis to bring out some important features of the finite element method freed, as we shall learn later, from the considerable complexity associated with the analysis of elements that are part of elastic continua.

In the next section we introduce the finite element method by applying it to an assembly of springs. To retain the greatest possible simplicity we shall choose a linear one-dimensional array.

4.1   THE DISPLACEMENTS IN A LINEAR ARRAY OF SPRINGS

4.1.1   THE BEHAVIOUR OF A SINGLE SPRING

Consider a coiled spring fixed at one end and with a force F applied at the other.

If the force required to produce unit extension of the spring is k (N mm^{-1}) then the displacement of the end of the spring u is given by the relation

$$F = ku$$

Consider now the spring in Fig. 4.1 that is part of a linear array. The nodes are labelled i and j and the spring element is identified by (e). The stiffness of the spring is $k^{(e)}$. We fix the right-hand node j and displace the left-hand node i by u_i.

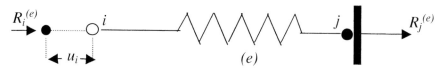

Fig. 4.1

We define forces to the right as positive. The reaction force at node i $R_i^{(e)}$ needed to keep i displaced is

$$R_i^{(e)} = +k^{(e)} u_i \qquad (4.1)$$

The force at j needed to keep the spring in equilibrium is given by the condition that the resultant force on the spring is zero. That is

$$R_i^{(e)} + R_j^{(e)} = 0 \quad \text{or} \quad R_j^{(e)} = -R_i^{(e)} = -k^{(e)} u_i \qquad (4.2)$$

Now fix node i and move node j by u_j (see Fig. 4.2)

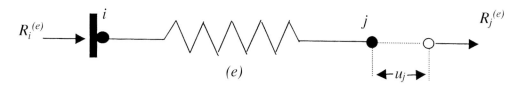

Fig. 4.2

The force $R_j^{(e)}$ needed to maintain the displacement u_j is

$$R_j^{(e)} = +k^{(e)} u_j \qquad (4.3)$$

The force at i needed to maintain the spring in equilibrium is given by

$$R_i^{(e)} + R_j^{(e)} = 0 \quad \text{or} \quad R_i^{(e)} = -R_j^{(e)} = -k^{(e)} u_j \qquad (4.4)$$

Now let both nodal displacements take place at the same time, then the force at i will be the sum of the forces due to the displacement at i, (u_i) and the displacement at j, (u_j). So from eqns (4.1) and (4.2)

$$R_i^{(e)} = k^{(e)} u_i - k^{(e)} u_j \qquad (4.5)$$

Similarly the force at node j will be, using eqns (4.3) and (4.4),

$$R_j^{(e)} = -k^{(e)} u_i + k^{(e)} u_j \qquad (4.6)$$

28

The two equations (4.5) and (4.6) for the forces at nodes i and j can be presented in matrix form:

$$\begin{Bmatrix} R_i^{(e)} \\ R_j^{(e)} \end{Bmatrix} = \begin{bmatrix} k^{(e)} & -k^{(e)} \\ -k^{(e)} & k^{(e)} \end{bmatrix} \begin{Bmatrix} u_i \\ u_j \end{Bmatrix} \qquad (4.7)$$

which has the general form

$$\{R\} = [K]\,\{U\} \qquad (4.8)$$

where $[K]$ is called the stiffness matrix of the spring element and is

$$[K] = \begin{bmatrix} k^{(e)} & -k^{(e)} \\ -k^{(e)} & k^{(e)} \end{bmatrix} \qquad (4.9)$$

Note that the matrix equation (4.8) has the form

Nodal forces = stiffness matrix × nodal displacements.

4.1.2 ANALYSIS OF A LINEAR ASSEMBLY OF SPRINGS

Now let us analyse the linear assembly of springs in Fig. 4.3.

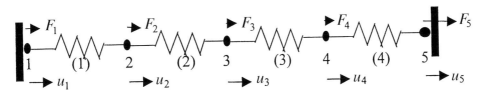

Fig. 4.3

The nodes are numbered 1 to 5 and the four spring elements, distinguished by brackets, are numbered from (1) to (4). The applied forces at the nodes are F_1 to F_5 and the nodal displacements are u_1 to u_5. Nodes 1 and 5 are fixed so $u_1 = u_5 = 0$.

Next we write down the condition that the forces at each node are in equilibrium. At node 1 the forces are the externally applied force F_1 and the internal reaction $R_1^{(1)}$. Note that, in the symbol for R, the suffix defines the node at which the force acts and the superscript defines the element whose distortion generates the force.

So equilibrium at node 1 demands that

$$R_1^{(1)} = F_1. \qquad (4.10)$$

At node 2 the internal reaction is the sum of that due to element 1 and element 2 that is

$$R_2^{(1)} + R_2^{(2)}$$

and this must equal the external force at node 2, F_2. So

$$R_2^{(1)} + R_2^{(2)} = F_2 \qquad (4.11)$$

29

By similar reasoning we get the conditions for force equilibrium at the other two nodes which are

$$R_3^{(2)} + R_3^{(3)} = F_3 \qquad (4.12)$$

$$R_4^{(3)} + R_4^{(4)} = F_4 \qquad (4.13)$$

$$R_5^{(4)} \qquad = F_5 \qquad (4.14)$$

We now relate the nodal reactions to the nodal displacements using the general eqs (4.5) and (4.6) which we derived earlier

For element (1)

$$R_1^{(1)} = k^{(1)}u_1 - k^{(1)}u_2$$

$$R_2^{(1)} = -k^{(1)}u_1 + k^{(1)}u_2$$

For element (2)

$$R_2^{(2)} = k^{(2)}u_2 - k^{(2)}u_3$$

$$R_3^{(2)} = -k^{(2)}u_2 + k^{(2)}u_3 \qquad (4.15)$$

For element (3)

$$R_3^{(3)} = k^{(3)}u_3 - k^{(3)}u_4$$

$$R_4^{(3)} = -k^{(3)}u_3 + k^{(3)}u_4$$

For element (4)

$$R_4^{(4)} = k^{(4)}u_4 - k^{(4)}u_5$$

$$R_5^{(4)} = -k^{(4)}u_4 + k^{(4)}u_5.$$

We can use this set of eqns (4.15) to express the force equilibrium eqns (4.10) to (4.14) in terms of the nodal displacements:

From (4.10) $\qquad (k^{(1)}u_1 - k^{(1)}u_2) \qquad\qquad\qquad = F_1$

From (4.11) $\qquad (-k^{(1)}u_1 + k^{(1)}u_2) + (k^{(2)}u_2 - k^{(2)}u_3) = F_2$

From (4.12) $\qquad (-k^{(2)}u_2 + k^{(2)}u_3) + (k^{(3)}u_3 - k^{(3)}u_4) = F_3$

From (4.13) $\qquad (-k^{(3)}u_3 + k^{(3)}u_4) + (k^{(4)}u_4 - k^{(4)}u_5) = F_4$

From (4.14) $\qquad (-k^{(4)}u_4 + k^{(4)}u_5) \qquad\qquad\qquad = F_5$

We can re-order these equations by collecting together the 'u' terms

$$
\begin{aligned}
k^{(1)}u_1 \quad -k^{(1)}u_2 \qquad\qquad\qquad\qquad\qquad &= F_1 \\
-k^{(1)}u_1 + (k^{(1)}+k^{(2)})u_2 \quad -k^{(2)}u_3 \qquad\qquad &= F_2 \\
-k^{(2)}u_2 + (k^{(2)}+k^{(3)})u_3 \quad -k^{(3)}u_4 \qquad &= F_3 \\
-k^{(3)}u_3 + (k^{(3)}+k^{(4)})u_4 - k^{(4)}u_5 &= F_4 \\
-k^{(4)}u_4 + k^{(4)}u_5 &= F_5
\end{aligned}
\qquad (4.16)
$$

Expressed in matrix format this is

$$
\begin{bmatrix}
k^{(1)} & -k^{(1)} & 0 & 0 & 0 \\
-k^{(1)} & k^{(1)}+k^{(2)} & -k^{(2)} & 0 & 0 \\
0 & -k^{(2)} & k^{(2)}+k^{(3)} & -k^{(3)} & 0 \\
0 & 0 & -k^{(3)} & k^{(3)}+k^{(4)} & -k^{(4)} \\
0 & 0 & 0 & -k^{(4)} & k^{(4)}
\end{bmatrix}
\begin{Bmatrix} u_1 \\ u_2 \\ u_3 \\ u_4 \\ u_5 \end{Bmatrix}
=
\begin{Bmatrix} F_1 \\ F_2 \\ F_3 \\ F_4 \\ F_5 \end{Bmatrix}
\tag{4.17}
$$

This has the form $[K]\{u\} = F$ where $[K]$ is the stiffness matrix of the spring array.

Equation (4.7) is the general stiffness matrix equation for a single spring. We can get the specific equation for element (1) of the spring assembly by putting the first two equations of (4.15) – those for element (1) – into the matrix form of (4.7) then

$$
\begin{Bmatrix} R_1^{(1)} \\ R_2^{(1)} \end{Bmatrix}
=
\begin{bmatrix} k^{(1)} & -k^{(1)} \\ -k^{(1)} & k^{(1)} \end{bmatrix}
=
\begin{Bmatrix} u_1 \\ u_2 \end{Bmatrix}
\tag{4.18}
$$

This is eqn (4.7) with $(e) = 1$ $i = 1$ and $j = 2$.

We can reproduce the stiffness matrix equation for the spring assembly, eqn (4.17), by combining the stiffness matrices of the individual spring elements. This process is called ASSEMBLY and is an important feature of the finite element method. We can appreciate the justification for this process if we copy the matrix equation for the spring array, eqn (4.17), and group sets of coefficients in the stiffness matrix.

$$
\begin{array}{ccccc}
u_1 & u_2 & u_3 & u_4 & u_5
\end{array}
$$
$$
\begin{bmatrix}
k^{(1)} & -k^{(1)} & 0 & 0 & 0 \\
-k^{(1)} & k^{(1)}+k^{(2)} & -k^{(2)} & 0 & 0 \\
0 & -k^{(2)} & k^{(2)}+k^{(3)} & -k^{(3)} & 0 \\
0 & 0 & -k^{(3)} & k^{(3)}+k^{(4)} & -k^{(4)} \\
0 & 0 & 0 & -k^{(4)} & k^{(4)}
\end{bmatrix}
\begin{Bmatrix} u_1 \\ u_2 \\ u_3 \\ u_4 \\ u_5 \end{Bmatrix}
=
\begin{Bmatrix} F_1 \\ F_2 \\ F_3 \\ F_4 \\ F_5 \end{Bmatrix}
\tag{4.19}
$$

The group of four stiffness coefficients in the dotted frame in rows 1 and 2 of (4.19) is the stiffness matrix of spring element (1), which is eqn (4.18). Similarly the group in rows 2 and 3 is the stiffness matrix for element (2), the group in rows 3 and 4 is that for element (3) and the group in rows 4 and 5 is that for element (4).

So we see that we can set up the stiffness matrix equation for the array of springs by assembling the stiffness matrices of the individual spring elements. We locate the stiffness coefficients in the appropriate row and column of the matrix for the total spring array. This matrix is called the GLOBAL MATRIX. To assist this process we have labelled the top of each column in (4.19) with the displacement degree of freedom that the coefficients in that column multiply. So, for example, to locate the coefficients for element (1), we note from Fig. 4.3 that the applied forces at the nodes of element (1) are F_1 and F_2 which appear in rows 1 and 2 of the global matrix. Also the matrix equation for element (1) in eqn (4.18) shows that the stiffness coefficients in the first column of the stiffness matrix multiply u_1 and therefore should be placed in column 1 of the global matrix, and in rows 1 and 2. The stiffness coefficients in the second column in (4.18) multiply u_2 and so should be located in the second column of the global matrix which has u_2 at its head. Proceeding in this way for each of the remaining three elements in turn

31

builds up the global matrix in a form identical to matrix equation (4.17) which we derived by developing the equations for the equilibrium of the forces at each node.

Mathematicians are exercised by the fact that the matrix equation we have derived in (4.17) is what they call singular. The determinant of the stiffness matrix is zero, so this means that when we try to invert the matrix equation, which involves dividing by the determinant, we multiply by infinity This creates a meaningless result which in mathematical terminology says that the inverse is not defined. The physical reason for this is that the spring array is floating in space and only becomes a problem amenable to analysis if we anchor at least one of the nodes. We shall see how this works out in the next section where we illustrate the application of the principles by solving a specific problem.

The final point to make in this section is to refer to the banded structure of the stiffness matrix in eqn (4.19). The second row of the matrix, for example, gives the force at node 2, F_2, which is the sum of three terms. We multiply out row 2 to get the equation for F_2

$$-k^{(1)}u_1 + (k^{(1)} + k^{(2)})u_2 - k^{(2)}u_3 = F_2$$

The first term $-k^{(1)}u_1$ is the force at node 2 due to a displacement of node 1 with node 2 and node 3 fixed. The second term $(k^{(1)} + k^{(2)})u_2$ is the force at node 2 due to a displacement at node 2 with node 1 and node 3 fixed. The third term $k^{(2)}u_3$ is the force at node 2 due to the displacement at node 3 with node 1 and node 2 fixed. No force is transmitted to node 2 when either node 4 or node 5 are displaced because during both of these translations nodes 1, 2 and 3 are fixed. So row 2 of the stiffness matrix has zeros in columns 4 and 5. The banded structure of the matrix can be understood by applying this principle to each row in turn.

4.1.3   LINEAR SPRING ARRAY

An Example Let us use the analysis in the previous section to determine the displacements in the following spring assembly.

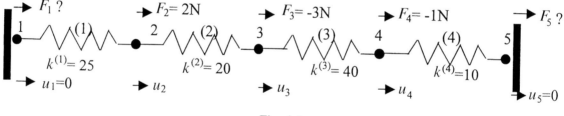

Fig. 4.4

The individual spring stiffnesses are given in Fig. 4.4 in $N\ mm^{-1}$. Nodes 1 and 5 are fixed so u_1 and u_5 are zero. The applied forces at nodes 2, 3 and 4 are indicated in newtons. We wish to determine the unknown displacements u_2, u_3 and u_4 and the unknown forces F_1 and F_5. We use the matrix eqn (4.19) and substitute the known spring stiffness coefficients and the known nodal forces. In this way we assemble the stiffness matrix for the spring array.

$$
\begin{bmatrix}
25 & -25 & 0 & 0 & 0 \\
-25 & (25+20) & -20 & 0 & 0 \\
0 & -20 & (20+40) & -40 & 0 \\
0 & 0 & -40 & (40+10) & -10 \\
0 & 0 & 0 & -10 & 10
\end{bmatrix}
\begin{Bmatrix}
u_1 \\ u_2 \\ u_3 \\ u_4 \\ u_5
\end{Bmatrix}
=
\begin{Bmatrix}
F_1 \\ 2 \\ -3 \\ -1 \\ F_5
\end{Bmatrix}
$$

We combine the terms in brackets

$$\begin{bmatrix} 25 & -25 & 0 & 0 & 0 \\ -25 & 45 & -20 & 0 & 0 \\ 0 & -20 & 60 & -40 & 0 \\ 0 & 0 & -40 & 50 & -10 \\ 0 & 0 & 0 & -10 & 10 \end{bmatrix} \begin{Bmatrix} 0 \\ u_2 \\ u_3 \\ u_4 \\ 0 \end{Bmatrix} = \begin{Bmatrix} F_1 \\ 2 \\ -3 \\ -1 \\ F_5 \end{Bmatrix} \qquad (4.20)$$

We have also substituted the conditions $u_1 = 0$ and $u_5 = 0$. Our first objective is to determine the three unknown nodal displacements u_2, u_3 and u_4. For this purpose we need only three equations. Rows 1 and 5 of (4.20) are equations for F_1 and F_5 which are two unknowns that are not our concern at the moment, so we can delete these two equations – that is rows 1 and 5. Also columns 1 and five multiply u_1 and u_5 respectively, but since both of these displacement terms are zero then all the products formed from the coefficients in columns 1 and 5 will also be zero so we can delete these two columns from the global stiffness matrix. The deleted rows and columns are indicated by dotted lines in (4.20).

The matrix formed by deleting these rows and columns is called the CONDENSED STIFFNESS MATRIX and is displayed in eqn (4.21). It has three unknowns and three equations and is therefore solvable.

$$\begin{bmatrix} 45 & -20 & 0 \\ -20 & 60 & -40 \\ 0 & -40 & 50 \end{bmatrix} \begin{Bmatrix} u_2 \\ u_3 \\ u_4 \end{Bmatrix} = \begin{Bmatrix} 2 \\ -3 \\ -1 \end{Bmatrix} \qquad (4.21)$$

We solve this for u_2, u_3 and u_4 using Gaussian reduction. First multiply row 2 by 45/20 so row 2 becomes

$$[-45 \quad 135 \quad -90]\{u\} = \{-6.75\}$$

now add row 1 to this and the matrix becomes

$$\begin{bmatrix} 45 & -20 & 0 \\ 0 & 115 & -90 \\ 0 & -40 & 50 \end{bmatrix} \begin{Bmatrix} u_2 \\ u_3 \\ u_4 \end{Bmatrix} = \begin{Bmatrix} 2 \\ -4.75 \\ -1 \end{Bmatrix}$$

multiply row 3 by 115/40 and row 3 becomes

$$[0 \quad -115 \quad 143.75]\{u\} = \{-2.875\}$$

Add row 2 to this and the matrix becomes

$$\begin{bmatrix} 45 & -20 & 0 \\ 0 & 115 & -90 \\ 0 & 0 & 53.75 \end{bmatrix} \begin{Bmatrix} u_2 \\ u_3 \\ u_4 \end{Bmatrix} = \begin{Bmatrix} 2 \\ -4.75 \\ -7.625 \end{Bmatrix}$$

Multiply out row 3

$$53.75u_4 = -7.625 \qquad u_4 = -0.14186 \text{ mm}$$

Multiply out row 2

$$115u_3 - 90u_4 = -4.75$$

substitute for u_4 and

$$u_3 = 1/115\{-4.75 + 90x(-0.14186)\} = -0.15232 \text{ mm}$$

Multiply out row 1

$$45u_2 - 20u_3 = 2$$

substitute for u_3 and

$$u_2 = 1/45\{20x(-0.15232) + 2\} = -0.02325 \text{ mm}$$

We now determine the two unknown forces F_1 and F_5 from the uncondensed matrix eqn (4.20). Expand row 1

$$25u_1 - 25u_2 = F_1 \quad \text{so} \quad F_1 = (25 \times 0)(25 \times [-0.02325])$$

and

$$F_1 = 0.5183 \text{ N}$$

Expand row 5

$$-10u_4 + 10u_5 = F_5 \quad \text{so} \quad F_5 = -(10 \times 0.14186) + (10 \times 0)$$

and

$$F_5 = 1.4186 \text{ N}$$

A useful check on the solution is that equilibrium demands that the resultant force on the array must be zero, that is

$$F_1 + F_2 + F_3 + F_4 + F_5 = 0$$

Check this:

$$0.5183 + 2 - 3 - 1 + 1.4186 = -0.0001 \text{ N} \qquad (4.22)$$

which is so close to zero to support the assertion that the solution is valid. The error in eqn (4.22) can be made smaller as more significant figures are taken in the calculations.

4.2 ANALYSIS OF A SIMPLE ONE-DIMENSIONAL CONTINUUM

We can readily extend the analysis of a linear array of springs to analyse simple one-dimensional continua where the stress is uniaxial. This will illustrate the first step in the analysis of elastic continua, which is the process of splitting the stressed body into finite elements – a process burdened with the inelegant title DISCRETEISATION. We are also concerned here with the relation between the forces applied to the element and the stresses generated. However in this element the force – stress relation is simple. The stress is the ratio of the axial force to the area of cross-section. Each element has a constant cross-section so the stress is uniform.

Consider an element in the form of a bar of uniform section. It has a length L, an area of cross-section A and elastic modulus E. We identify the element as (e) and label the nodes i and j

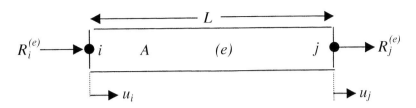

We can treat this element in the same way that we analysed the single spring. Fix node j and displace node i by u_i. The reaction at i needed to keep i displaced is $R_i^{(e)}$. This force will generate a stress σ given by

$$R_i^{(e)} = A\sigma = AE\varepsilon = AE\frac{u_i}{L} \quad \text{where } \varepsilon \text{ is the axial strain in the element.}$$

By analogy with the spring

$$R_i^{(e)} = \frac{AE}{L}u_i = k^{(e)}u_i \text{ where } k^{(e)} = \frac{AE}{L} \quad \text{is the stiffness of the bar.}$$

The reaction of the bar to applied forces is similar in essence to that of the single spring that we analysed in the previous section. The stiffness matrix of the bar element will be similar to that of eqn (4.7) for the spring element with $k^{(e)}$ replaced by $\frac{AE}{L}$.

So the stiffness matrix of the bar element is

$$\left\{ \begin{array}{c} R_i^{(e)} \\ R_j^{(e)} \end{array} \right\} = \begin{bmatrix} \dfrac{AE}{L} & \dfrac{-AE}{L} \\ \dfrac{-AE}{L} & \dfrac{AE}{L} \end{bmatrix} \left\{ \begin{array}{c} u_i \\ u_j \end{array} \right\} \tag{4.23}$$

As a simple illustration of the application of this element stiffness matrix to solve a real problem we work out the overall extension and the stress distribution in a linearly tapered bar of circular section with an end load. For variety we again use Imperial units for this exercise.

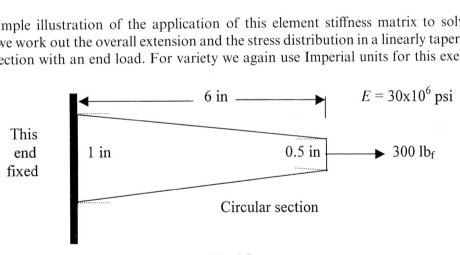

Fig. 4.5

We divide the bar into three elements of equal length 2 in and make the diameter of each element the mean of the two end diameters of the element. The element mesh is drawn in Fig. 4.6 where the elements and the nodes are numbered.

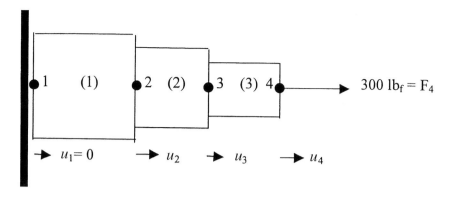

Fig. 4.6

The element areas are: $A_{(1)} = 0.6599$ in^2 $A_{(2)} = 0.4418$ in^2 and $A_{(3)} = 0.2672$ in^2. The element stiffness coefficients , $\frac{AE}{L}$, are

$$k^{(1)} = \frac{0.6599E}{L} \quad k^{(2)} = \frac{0.4418E}{L} \quad k^{(3)} = \frac{0.2672E}{L}$$

We now assemble the stiffness matrices of the individual elements into the global stiffness matrix for the whole model.

$$
\begin{array}{cccc}
u_1 & u_2 & u_3 & u_4
\end{array}
$$

$$
\begin{bmatrix}
k^{(1)} & -k^{(1)} & 0 & 0 \\
-k^{(1)} & k^{(1)} & 0 & 0 \\
0 & k^{(2)} & -k^{(2)} & 0 \\
0 & -k^{(2)} & k^{(2)} & 0 \\
0 & 0 & k^{(3)} & -k^{(3)} \\
0 & 0 & -k^{(3)} & k^{(3)}
\end{bmatrix}
\begin{Bmatrix}
u_1 \\ u_2 \\ u_3 \\ u_4
\end{Bmatrix}
=
\begin{Bmatrix}
F_1 \\ F_2 \\ F_3 \\ F_4
\end{Bmatrix}
\qquad (4.24)
$$

The four force equations have been separated by horizontal dotted lines in the matrix to emphasise, for example, that the equation for F_2 is formed by adding together the stiffness coefficients in the columns of rows 2 and 3. Again we have headed each column with the degree of freedom which the coefficients in the column multiply when the force equation is multiplied out.

We now substitute the numerical values of the stiffness coefficients in (4.24) and take outside the matrix the common factor $E/2$. We also substitute the known applied nodal forces.

$$
\frac{E}{2}
\begin{array}{cccc}
u_1 & u_2 & u_3 & u_4
\end{array}
$$

$$
\frac{E}{2}
\begin{bmatrix}
.6599 & -.6599 & 0 & 0 \\
-.6599 & .6599 & 0 & 0 \\
0 & .4418 & -.4418 & 0 \\
0 & -.4418 & .4418 & 0 \\
0 & 0 & .2672 & -.2672 \\
0 & 0 & -.2672 & .2672
\end{bmatrix}
\begin{Bmatrix}
u_1 \\ u_2 \\ u_3 \\ u_4
\end{Bmatrix}
=
\begin{Bmatrix}
F_1 \\ 0 \\ 0 \\ 300
\end{Bmatrix}
\qquad (4.25)
$$

36

Adding up the stiffness coefficients between the dotted lines we get the global stiffness matrix in a form that we can process:

$$\frac{E}{2}\begin{bmatrix} 0.6599 & -0.6599 & 0 & 0 \\ -0.6599 & 1.1017 & -.4418 & 0 \\ 0 & -0.4418 & 0.7090 & -0.2672 \\ 0 & 0 & -0.2672 & 0.2672 \end{bmatrix}\begin{Bmatrix} u_1 \\ u_2 \\ u_3 \\ u_4 \end{Bmatrix} = \begin{Bmatrix} F_1 \\ 0 \\ 0 \\ 300 \end{Bmatrix} \qquad (4.26)$$

The first row of (4.25) is the equation for F_1 which is unknown so we delete this equation at this stage by eliminating row 1. Also, since the left hand end of the tapered bar is fixed, $u_1 = 0$ so the terms which multiplying this displacement, which are all the coefficients in column 1, are zero so we lose nothing by eliminating column. These two deletions give us the CONDENSED MATRIX which follows

$$\frac{E}{2}\begin{bmatrix} 1.1017 & -0.4418 & 0 \\ -0.4418 & 0.7090 & -0.2672 \\ 0 & -.2672 & 0.2672 \end{bmatrix}\begin{Bmatrix} u_2 \\ u_3 \\ u_4 \end{Bmatrix} = \begin{Bmatrix} 0 \\ 0 \\ 300 \end{Bmatrix} \qquad (4.27)$$

Note that the stiffness matrix is symmetric about the leading diagonal; (4.27) contains three equations and three unknowns, u_2 u_3 and u_4 so there is adequate information to obtain the three displacements. We will use Gaussian reduction, which we introduced in Chapter 3, to get a solution.

First we multiply row 1 by 0.4418/1.107. Row 1 becomes (ignoring the $E/2$)

$$[-0.4418 \quad -0.1772 \quad 0 \quad]\{ \} = \{0\}$$

Now add this to row 2 and the matrix becomes

$$\frac{E}{2}\begin{bmatrix} 1.107 & -0.4418 & 0 \\ 0 & 0.5318 & -0.2672 \\ 0 & -0.2672 & 0.2672 \end{bmatrix}\begin{Bmatrix} u_2 \\ u_3 \\ u_4 \end{Bmatrix} = \begin{Bmatrix} 0 \\ 0 \\ 300 \end{Bmatrix}$$

Multiply row 2 by 0.2672/0.5318 and row 2 becomes (without the $E/2$)

$$[0 \quad 0.2672 \quad -0.1343]\{ \} = \{0\}$$

Add this to row 3 and the matrix is in its desired form with all the terms below the leading diagonal zero:

$$\frac{E}{2}\begin{bmatrix} 1.107 & -.4418 & 0 \\ 0 & 0.5318 & -0.2672 \\ 0 & 0 & 0.1329 \end{bmatrix}\begin{Bmatrix} u_2 \\ u_3 \\ u_4 \end{Bmatrix} = \begin{Bmatrix} 0 \\ 0 \\ 300 \end{Bmatrix}$$

Multiply out row 3

$$E/2 \times 0.1329\, u_4 = 300 \quad \text{so} \quad u_4 = \frac{2 \times 300}{0.1329 \times 30 \times 10^6} \quad u_4 = 1.5049 \times 10^{-4} \text{ in}$$

Multiply out row 2

$$E/2(0.5318\,u_3 - 0.2672\,u_4) = 0 \quad \text{substitute } u_4 \text{ and } u_3 = 7.561 \times 10^{-5} \text{ in}$$

Multiply out row 1

$$E/2(1.1071\,u_2 - 0.4418\,u_3) = 0 \text{ substitute } u_3 \text{ and } u_2 = 3.032 \times 10^{-5} \text{ in}$$

We can now work out the axial stress in each element

$$\sigma = E\varepsilon = E\left(\frac{\{u_j - u_i\}}{L}\right)$$

For element (1)

$$\sigma_{(1)} = E/2(u_2 - 0) = 454.8 \text{ psi } (454.6)$$

For element (2)

$$\sigma_{(2)} = E/2(u_3 - u_2) = 679.3 \text{ psi } (679.0)$$

For element (3)

$$\sigma_{(3)} = E/2(u_4 - u_3) = 1123.2 \text{ psi } (1122.7)$$

The actual stress is simply obtained by dividing the end load by the element area. The values are given in brackets after the finite element value. The latter are seen to be close to the exact values. The overall extension of the tapered bar u_4, which is the finite element result, is

$$1.5049 \times 10^{-4} \text{ in}$$

The exact extension can be obtained by integrating the strain along the bar

$$\text{Extension} = \int_0^6 \varepsilon.dx = \int_0^6 \frac{\sigma}{E}.dx = \int_0^6 \frac{300}{AE}.dx$$

The diameter of the bar at a distance x from the fixed end is

$$\left(1 - x\frac{0.5}{6}\right) \text{ so the area is } \frac{\pi}{4}\left(1 - \frac{x}{12}\right)^2$$

and the integral becomes

$$\int_0^6 \frac{4L}{\pi E\left\{1 - \frac{x}{12}\right\}^2}.dx = \frac{4 \times 300}{\pi \times 30 \times 10^6}\left[\frac{12}{1 - \frac{x}{12}}\right]_0^6 = 1.5279 \times 10^4 \text{ in.}$$

The finite element solution is only 1.5% less than this so it is a quite good estimate.

The finite element method always yields approximate solutions. The error diminishes as the number of elements in the model increases. This can be demonstrated by working through the

38

tapered bar example with a different number of elements. The result of such calculations for the overall extension of the bar, the details of which we omit, is:

Number of elements	Overall extension (in)
1	1.358×10^{-4}
2	1.476×10^{-4}
3	1.505×10^{-4}
10	1.525×10^{-4}
∞	1.528×10^{-4}

As the number of elements becomes large, the finite element solution converges on the exact solution.

This example, of course, can be worked out exactly by classical analysis and as such it is not really a candidate for finite element analysis. But, because it is so simple, it brings out some of the features of the finite element method clearly, and because an exact solution is available it is possible to gauge exactly the degree of error in the finite element solution.

CHAPTER 5   ENERGY PRINCIPLES IN FINITE ELEMENT ANALYSIS

We shall now turn our attention to the analysis of the stress distribution in elastic continua. These are bodies in which the stress varies continuously from point to point as, for example, in the loaded elastic sheet with a central hole that we presented in Chapter 1. In the next chapter we will develop the analysis based on a division of the loaded body into triangular elements.

In the arrays of springs or of bars under axial load that have been our concern so far it has been possible to determine the stiffness coefficients in the matrix quite simply. Further, in the case of the bar elements, the relation between the load and the stress in the bar is also simple. When we come to look at triangular continuum elements we shall find that both of the tasks that are simple for the one-dimensional elements turn out to be much more complex for the elements that are part of an elastic body, so much so that we need additional equations to achieve solutions. These are secured by applying energy principles, so before getting involved in continua elements it is necessary to explain the energy principles and their origin in some detail. We shall begin with:

5.1   THE PRINCIPLE OF VIRTUAL WORK

Consider a number of forces acting at a point P. The forces are in equilibrium, which means that the resultant of all the components of the forces resolved in any direction is zero.

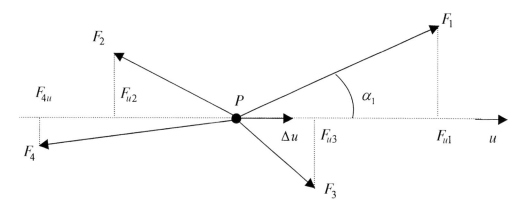

Fig. 5.1

In Fig. 5.1 resolve the four forces F_1, F_2, F_3 and F_4 in the horizontal direction u. F_{u1} is the horizontal component of F_1 in that direction and is $F_1 \cos \alpha_1$ and similarly for the other horizontal components F_{u2}, F_{u3} and F_{u4}.

The equilibrium condition that the sum of these horizontal components is zero is

$$F_{u1} + F_{u2} + F_{u3} + F_{u4} = 0 \tag{5.1}$$

Now give P a very small displacement Δu in the horizontal direction of u. This displacement is assumed to be so small that the forces are unchanged by it. We call such a displacement a VIRTUAL DISPLACEMENT. The work done by each of the four forces during this displacement is the force x displacement because the forces do not change, so the total work done is

$$V_u = F_{u1}\Delta u + F_{u2}\Delta u + F_{u3}\Delta u + F_{u4}\Delta u$$

or $\quad V_u = (F_{u1} + F_{u2} + F_{u3} + F_{u4})\Delta u$ $\qquad$ (5.2)

But the condition for equilibrium, eqn (5.1), has not been violated by the virtual displacement because the forces are unchanged, so

$$(F_{u1} + F_{u2} + F_{u3} + F_{u4}) = 0 \quad \text{in eqn (5.2)}$$

therefore $V_u = 0$, which means that the total work done during the virtual displacement is zero. The VIRTUAL WORK PRINCIPLE states that

> A necessary and sufficient condition for equilibrium of a particle is that the work done by forces on the particle is zero under any virtual displacement.

A very useful extension of the principle which we shall use in the development of the finite element equations is the following:

Consider a particle I attached to four other particles by springs, through which the particles can exert a force on each other (Fig. 5.2).

Fix particles 1, 2, 3 and 4 and apply a force F_I to the particle I in the direction shown in Fig. 5.2. This generates tensile forces in the springs F_{I1}, F_{I2}, F_{I3} and F_{I4}.

Now displace I in the direction of F_I by a small virtual displacement Δu_I. This virtual displacement leaves the tensile forces in the four springs unchanged because it is so small. By the Principle of Virtual Work, since I is in equilibrium, the total work done by all the forces acting on it

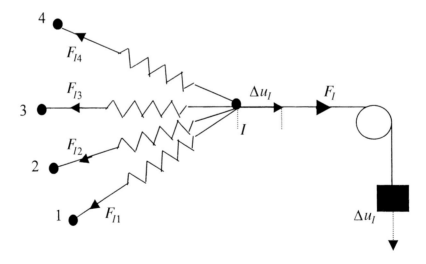

Fig. 5.2

is zero. The energy stored in the springs as the result of the displacement Δu_I will be

$$\Delta U = F_{I1}\Delta u_{I1} + F_{I2}\Delta u_{I2} + F_{I3}\Delta u_{I3} + F_{I4}\Delta u_{I4} \qquad (5.3)$$

where Δu_{I1} is the extension of the spring I-1 resulting from the displacement Δu_I of I, and correspondingly for the other three springs. All the energy terms will be positive because the virtual displacement increases the elastic energy in the springs. Note that the work done is force x displacement; there is no factor $\frac{1}{2}$ which arises in situations where the force increases linearly with the displacement.

We now need to work out the change of potential energy when the applied force is displaced by Δu_I during the virtual displacement. This change can be understood if we imagine that the force F_I is applied by a string which passes over a frictionless pulley and carries a weight W, where

$$W = \frac{F_I}{g} \text{ and } g \text{ is the gravitational acceleration.}$$

When I is displaced by Δu_I, the weight will fall by the same distance and the its gravitational potential will change by

$$\Delta V = -Wg\Delta u_I = -F_I \Delta u_I \tag{5.4}$$

So ΔV is the change of potential energy of the applied force and is negative because it is a decrease.

The Principle of Virtual Work requires that, since I is in equilibrium, the change in the total energy of the system during the virtual displacement is zero so

$$\Delta U + \Delta V = 0 \tag{5.5}$$

so

$$(F_{I1}\Delta u_{I1} + F_{I2}\Delta u_{I2} + F_{I3}\Delta u_{I3} + F_{I4}\Delta u_{I4}) - F_I \Delta u_I = 0 \tag{5.6}$$

Equation (5.5) gives

$$\Delta U = -\Delta V \tag{5.7}$$

That is (change of elastic strain energy in the elastic members) = (reduction of potential energy of the applied load) during a virtual displacement.

We shall see in the next chapter that this principle is used to relate the nodal forces on an element to the stresses within the element. Referring to Fig. 5.2, if 1, 2, 3, 4 and I are nodes in a finite element model, then if we fix nodes 1, 2, 3 and 4 and then give I a small virtual displacement we can use the relation in eqn (5.5) to relate the nodal load to the strain energy in the elastic elements which in the case in Fig. 5.2 comprise a set of springs. If the element is a triangular element in an elastic continuum, then the strain energy can be related to the element stresses. So in these circumstances the virtual work principle makes it possible to relate nodal forces to element stresses. We shall see how this operates in practice.

An alternative approach to the Principle of Virtual Work and very closely related to it is the Principle of Stationary Total Potential. Mathematicians refer to this as a Variational Principle. It achieves the same objective in finite element analysis as does the Virtual Work Principle in that it yields the relationship between nodal forces and nodal stresses. However it has a more abstract mathematical feel about it than does the Virtual Work principle and it is for this reason we have chosen to use the latter. Nevertheless variational principles are important in the analytical basis of the finite element method so it is helful to say a little on the subject.

5.2 THE PRINCIPLE OF STATIONARY TOTAL POTENTIAL

Equation (5.5) can be expressed as

$$\Delta(U + V) = 0 \tag{5.8}$$

Where $(U + V)$ is the sum of the elastic stored energy in the elements (U) and the potential energy of the nodal forces (V). This sum is the total potential energy which we shall call χ. So (5.8) states that

$$\Delta\chi = 0 \tag{5.9}$$

We recall that (5.7) and the related (5.8) are statements of the condition for equilibrium of a particle subjected to applied forces so (5.9) implies that in equilibrium the change of total potential energy if conditions are shifted slightly is zero. This is equivalent mathematically to the statement that χ is at a minimum, because at the minimum of any function, a small change of any of the independent variables leaves the dependent variable unchanged – or stationary.

So the equation $\Delta\chi = 0$ is a statement that, in equilibrium, the total potential energy is a minimum. This statement is the Principle of Stationary Total Potential.

In finite element analysis, χ is called a FUNCTIONAL and the differentiation of the functional to find the minimum value is an alternative to the Principle of Virtual Work for relating nodal forces to element stresses. A functional is a function of another function, and it turns out that χ is of that ilk.

Let us pause here to remind ourselves of the calculus of maxima and minima. We know that if we have a relation

$$y = x^2 + 3x + 2$$

then to find a minimum value we put

$$\frac{dy}{dx} = 0$$

this yields $2x + 3 = 0$ or $x = -1.5$. To satisfy ourselves that this is a minimum we also need to check that

$$\frac{d^2y}{dx^2}$$

is positive. It is $+2$.

The rationale behind this bit of calculus can be grasped in the following way. Consider the graph in Fig. 5.3.

Consider the point A on the curve. Displace the point a small distance Δx in the x direction. Then A will move up the curve Δy in the y direction. Now Δy will be quite close to $\Delta x \tan\alpha$, where α is the angle that the tangent to the curve at A makes with the x-axis. The error will become smaller as Δx gets smaller. Now $\tan\alpha$ is the gradient of the tangent at A which is $\frac{dy}{dx}$ so

$$\Delta y = \Delta x \tan\alpha = \Delta x \frac{dy}{dx} \tag{5.10}$$

If we now move to the point B on the curve at the minimum, $\frac{dy}{dx} = 0$ so a very small movement Δx

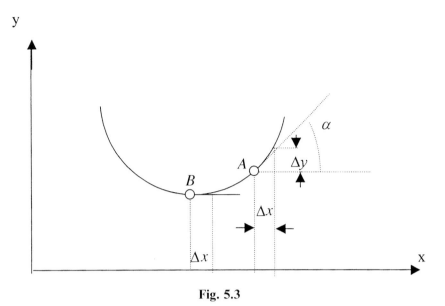

Fig. 5.3

from here along the x-axis will cause y to change by

$$\Delta y = \Delta x \frac{dy}{dx} = \Delta x.0 = 0$$

So at the minimum a small change of x causes no change of y.

Figure 5.3 is concerned with only a single independent variable x. If there were two independent variables as would be the case if we had a function z that was a function of x and y, then the three dimensional plot of z would be a surface. If that surface were bowl-shaped then at the bottom of the bowl, which would be a two-dimensional minimum, a small movement in either the x or the y direction would produce no change in z. In analytical terms this means that

$$\Delta z = \frac{\partial z}{\partial x} \Delta x = 0 \quad \text{and} \quad \Delta z = \frac{\partial z}{\partial y} \Delta y = 0.$$

We use curly ∂ in the differentials to indicate that, for example in $\frac{\partial z}{\partial x}$, x is varied whilst all other dependent variables are kept constant.

In finite element analysis we meet variables that are functions of many independent variables. For example the displacement at a point within a two-dimensional continuum element with four nodes will be a function of eight nodal displacements – two at each node. These displacements are called Degrees of Freedom. The principles we have outlined still apply to this more complex system.

We saw in Fig. 5.3 and in eqn (5.10) that if we change x by Δx at A then y changes by

$$\Delta y = \frac{dy}{dx} \Delta x \tag{5.11}$$

(we have reversed the order of the terms here compared with eqn (5.10) to be consistent with the usage that follows). Now with a function of several variables, a relation like (5.11) applies to changes of each variable in turn, but we must recall that when one independent variable is changed all the others are kept constant. We achieve this by using partial differentiation with curly ∂s.

If we now return to our Total Potential χ which, for example, will be a function of a number of nodal displacements u_1, u_2, u_3, u_4 and so on, then if in turn u_1 is changed by Δu_1, u_2 is changed by Δu_2 and so on, the overall change in χ is

$$\Delta\chi = \frac{\partial\chi}{\partial u_1}\Delta u_1 + \frac{\partial\chi}{\partial u_2}\Delta u_2 + \frac{\partial\chi}{\partial u_3}\Delta u_3 + \frac{\partial\chi}{\partial u_4}\Delta u_4 + \ldots\ldots\ldots\ldots \quad (5.12)$$

Now at the multi-dimensional minimum all the partial derivatives are zero – that is movement in any of the dimensions with all the others fixed does not change χ. So at a minimum

$$\frac{\partial\chi}{\partial u_1} = 0 \quad \frac{\partial\chi}{\partial u_2} = 0 \quad \frac{\partial\chi}{\partial u_3} = 0 \quad \frac{\partial\chi}{\partial u_4} = 0 \ldots\ldots\ldots\ldots \frac{\partial\chi}{\partial u_n} = 0 \quad (5.13)$$

for an assembly with n degrees of freedom.

We have seen that the condition for equilibrium is that $\Delta\chi = 0$ [eqn (5.9)] or χ is a minimum. This is identical to the condition leading to eqn (5.13), so (5.13) is the analytical statement of the equilibrium state which is the Condition for Stationary Total Potential. Equation (5.13) can be used as an alternative to that derived from the Virtual Work Principle to solve finite element problems.

We illustrate this by working through the problem we addressed in Chapter 4 using both energy methods.

5.3   AN ILLUSTRATIVE EXAMPLE

We determine the three unknown displacements and the two unknown forces in the linear array of springs of Fig. 4.4. We will first present it in the general form of Fig. 5.4.

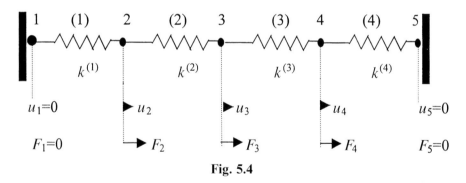

Fig. 5.4

The four spring elements and the five nodes are numbered. Nodes 1 and 5 are fixed. Forces F_2 F_3 and F_4 are applied to nodes 2, 3 and 4 where the nodal displacements are u_2, u_3 and u_4.

5.3.1   Apply the Principle of Virtual Work

First give node 2 a virtual displacement Δu_2 keeping nodes 3 and 4 fixed. Note that nodes 1 and 5 are permanently fixed. Now apply the principle that the reduction of the potential energy of the force at node 2, F_2 is equal to the increase of elastic energy in springs (1) and (2). Springs (3) and (4) are unaffected by the displacement Δu_2 because nodes 3, 4 and 5 are fixed so neither spring changes length.

Let the tensions in springs (1) and (2) be T_1 and T_2.

$$T_1 = k^{(1)}(u_2 - u_1)$$

where $k^{(1)}$ is the spring stiffness and $(u_2 - u_1)$ is the spring extension. But

$$u_1 = 0 \text{ so } T_1 = k^{(1)}u_2 \qquad (5.14)$$

Recall that u_2 is very much greater than the virtual displacement so T_1 is the same before and after the displacement. Similarly

$$T_2 = k^{(2)}(u_3 - u_2) \qquad (5.15)$$

The energy stored in the two springs as a result of the virtual displacement Δu_2 is

$$T_1 \Delta u_2 - T_2 \Delta u_2 \qquad (5.16)$$

The second term is negative because the virtual displacement reduces the extension of spring (2). Substitute (5.14) and (5.15) in (5.16) then the elastic stored energy becomes

$$k^{(1)}u_2 \Delta u_2 - k^{(2)}(u_3 - u_2)\Delta u_2 \qquad (5.17)$$

The fall in potential energy of F_2 is $F_2 \Delta u_2$ so the energy balance is [eqn (5.7)]

$$k^{(1)}u_2 \Delta u_2 - k^{(2)}(u_3 - u_2)\Delta u_2 = F_2 \Delta u_2$$

And Δu_2 cancels on the two sides of the equation so

$$k^{(1)}u_2 - k^{(2)}(u_3 - u_2) = F_2 \qquad (5.18)$$

Similarly for springs (2) and (3) with nodes 2 and 4 fixed and node 3 given a virtual displacement Δu_3 and T_3 the tension in element (3) then

$$T_2 \Delta u_3 - T_3 \Delta u_3 = F_3 \Delta u_3$$

and Δu_3 cancels so substituting for T_2 and T_3

$$k^{(2)}(u_3 - u_2) - k^{(3)}(u_4 - u_3) = F_3 \qquad (5.19)$$

and fixing nodes 2 and 3 and displacing node 4 by a virtual displacement Δu_4 gives, where T_4 is the tension in element (4),

$$T_3 \Delta u_4 - T_4 \Delta u_4 = F_4 \Delta u_4$$

and Δu_4 cancels and relating the spring tensions to the spring extensions

$$k^{(3)}(u_4 - u_3) - k^{(4)}(u_5 - u_4) = F_4$$

but $u_5 = 0$ so

$$k^{(3)}(u_4 - u_3) + k^{(4)}u_4 = F_4 \qquad (5.20)$$

If we gather together like terms in u in equations (5.18), (5.19) and (5.20) we get the following three equations that relate the nodal displacements to the applied nodal forces

$$
\begin{aligned}
(k^{(1)} + k^{(2)})u_2 \quad\quad -k^{(2)}u_3 \quad\quad\quad\quad &= F_2 \\
-k^{(2)}u_2 + (k^{(2)} + k^{(3)})u_3 \quad\quad -k^{(3)}u_4 &= F_3 \\
-k^{(3)}u_3 + (k^{(3)} + k^{(4)})u_4 &= F_4
\end{aligned}
$$

These three equations can conveniently be presented in compact matrix form

$$
\begin{bmatrix}
(k^{(1)} + k^{(2)}) & -k^{(2)} & 0 \\
-k^{(?)} & (k^{(2)} + k^{(3)}) & -k^{(3)} \\
0 & -k^{(3)} & (k^{(3)} + k^{(4)})
\end{bmatrix}
\begin{Bmatrix} u_2 \\ u_3 \\ u_4 \end{Bmatrix}
=
\begin{Bmatrix} F_2 \\ F_3 \\ F_4 \end{Bmatrix}
\tag{5.21}
$$

We put in (5.21) the particular values for the problem with four springs which we analysed in Chapter 4. The details are given in Fig. 4.4. They are

$$F_2 = 2\text{N} \quad F_3 = -3\text{N} \quad F_4 = -1\text{N}$$

$$K^{(1)} = 25 \text{ N/mm} \quad k^{(2)} = 20 \text{ N/mm} \quad k^{(3)} = 40 \text{ N/mm} \quad k^{(4)} = 10 \text{ N/mm}$$

$$
\begin{bmatrix}
45 & -20 & 0 \\
-20 & 60 & -40 \\
0 & -40 & 50
\end{bmatrix}
\begin{Bmatrix} u_2 \\ u_3 \\ u_4 \end{Bmatrix}
=
\begin{Bmatrix} 2 \\ -3 \\ -1 \end{Bmatrix}
\tag{5.22}
$$

This is identical to the matrix eqn (4.21) derived in Chapter 4 for the same problem solved by writing down the force equilibrium conditions at each node. So eqn (5.22) will give the same values for the nodal displacements u_2, u_3 and u_4 as did equation (4.21).

It is timely to recall at the end of this section our discussion in Chapter 3 on the significance of the coefficients in the stiffness matrix. We made the point that in the analysis of load transfer by nodal displacement all nodes are fixed aside from the active node. It is in the application of the Principle of Virtual Work in this section that we see this condition imposed in the analysis.

To complete the comparisons we now solve the same problem by minimising the Total Potential.

5.3.2 APPLY THE PRINCIPLE OF MINIMUM TOTAL POTENTIAL

First we develop an expression for the strain energy stored in a loaded spring element. Consider a single spring

Fig. 5.5

If the ends are displaced by u_i and u_j the spring extension is $(u_j - u_i)$ and if the tension developed in the spring is T then the elastic stored energy is $\frac{1}{2}$ tension x extension $= \frac{1}{2}T(u_j - u_i)$.

The $\frac{1}{2}$ arises because the tension increases linearly with the extension. If the spring stiffness is k then $T = k(u_j - u_i)$ and, substituting this for T, the stored energy is

$$\tfrac{1}{2}k(u_j - u_i)^2 \tag{5.23}$$

Addressing now the spring array in Fig. 5.4, the total elastic strain energy in the four springs will be the sum of four terms like (5.23). Recalling that u_1 and u_5 are both zero, the strain energy is

$$\tfrac{1}{2}k^{(1)}u_2^2 + \tfrac{1}{2}k^{(2)}(u_3 - u_2)^2 + \tfrac{1}{2}k^{(3)}(u_4 - u_3)^2 + \tfrac{1}{2}k^{(4)}u_4^2.$$

The reduction of the potential energy of the applied forces during the nodal displacements is

$$F_2 u_2 + F_3 u_3 + F_4 u_4.$$

So the total potential energy of the spring array after loading is

$$\chi = \tfrac{1}{2}k^{(1)}u_2^2 + \tfrac{1}{2}k^{(2)}(u_3 - u_2)^2 + \tfrac{1}{2}k^{(3)}(u_4 - u_3)^2 + \tfrac{1}{2}k^{(4)}u_4^2 - F_2 u_2 - F_3 u_3 - F_4 u_4 \tag{5.24}$$

Now apply the condition that χ must not change if any node is given a small displacement with all the other nodes fixed. Taking node 2 first, this condition is

$$\frac{\partial \chi}{\partial u_2} = 0$$

The curly ∂ implies that the derivative is giving the rate of change of χ with u_2 under conditions where all other variables are constrained not to change. So in obtaining the partial derivative we assume all u variables other than u_2 are constant, then from (5.24)

$$\frac{\partial \chi}{\partial u_2} = k^{(1)}u_2 - k^{(2)}(u_3 - u_2) - F_2 = 0$$

gathering like u terms together and re arranging

$$(k^{(1)} + k^{(2)})u_2 - k^{(2)}u_3 = F_2 \tag{5.25}$$

Now move node 3 with nodes 2 and 4 fixed then, again from (5.24)

$$\frac{\partial \chi}{\partial u_3} = k^{(2)}(u_3 - u_2) - k^{(3)}(u_4 - u_3) - F_3 = 0$$

or

$$-k^{(2)}u_2 + (k^{(2)} + k^{(3)})u_3 - k^{(3)}u_4 = F_3 \tag{5.26}$$

Now move node 4 with nodes 2 and 3 fixed then from (5.24)

$$\frac{\partial \chi}{\partial u_4} = k^{(3)}(u_4 - u_3) + k^{(4)}u_4 - F_4 = 0$$

or

$$-k^{(3)}u_3 + (k^{(3)} + k^{(4)})u_4 = F_4 \tag{5.27}$$

Gathering the three equations (5.25), (5.26) and (5.27) into matrix form

$$
\begin{bmatrix}
(k^{(1)} + k^{(2)}) & -k^{(2)} & 0 \\
-k^{(2)} & (k^{(2)} + k^{(3)}) & -k^{(3)} \\
0 & -k^{(3)} & (k^{(3)} + k^{(4)})
\end{bmatrix}
\begin{Bmatrix} u_2 \\ u_3 \\ u_4 \end{Bmatrix}
=
\begin{Bmatrix} F_2 \\ F_3 \\ F_4 \end{Bmatrix}
\tag{5.28}
$$

This is the identical matrix equation to eqn (5.21) derived using the Principle of Virtual Work to solve the same problem. This demonstrates that the two principles provide alternative ways of solving these problems. It is worth commenting too that the process of equating the partial derivatives of χ to zero in this analysis implies that all nodes other than one are fixed during this change. This emphasises again our discussion in Chapter 3 that the coefficients in the stiffness matrix are derived on this assumption.

We have now covered all the basic ideas that we need to address the finite element analysis of loaded elastic continua. In the next chapter we begin by analysing the behaviour of the simplest element used for this analysis, referred to as the Constant Strain Triangle.

CHAPTER 6 FINITE ELEMENTS THAT FORM PART OF A CONTINUUM

So far we have analysed structures that are assemblies of elements that are either springs or pin-jointed rods. In these elements the relationship between the nodal forces and the tension or stress in the element is particularly simple. We now move to the more complex problem of the analysis of two-dimensional elements that form part of an elastic continuum, like the loaded sheet with a circular hole which we presented in Chapter 1. In order to work out the stress distribution in such a loaded body, we divide it into elements which are connected at nodes at their corners (in the first instance). We work out the relationship between the forces applied to the individual elements and the nodal displacements. which involves setting up the element stiffness matrices. We then assemble the element stiffness matrices, in exactly the same way as we did in matrix eqn (4.19) of Chapter 4 for the linear array of four springs, and get the global stiffness matrix for the whole element assembly. This relates all the applied forces to the nodal displacements. Knowing the applied forces, we can invert the stiffness matrix and get the nodal displacements. From these, it turns out, we can derive the elastic strains in each element and from these the stresses. In this way we derive the stress distribution in the whole loaded body.

The analysis is basically more complex than that which has sufficed to solve the problems we have addressed so far, because the relationship between the forces applied to a continuum element and the strains and stresses that these forces generate is far from trivial. We shall see that we need to deploy the energy concepts that we developed in Chapter 5 to establish this force-stress relationship.

Let us assume that we wished to determine the stress distribution in the loaded elastic sheet with a central circular hole which we introduced in Chapter 1. This, of course, is an important problem in the understanding of material behaviour particularly under fatigue stressing because the analysis will quantify the stress concentration factor introduced by the presence of the hole.

We begin by dividing the body into a mesh of elements. This is called discretisation in finite element jargon. There is a large range of different types of element which are listed in the handbooks of proprietary finite element packages. In this chapter we shall choose the triangular element with uniform strain. In the next chapter we will introduce the more versatile quadrilateral element in which the strain varies across it.

The pattern of elements in which the stressed body is arbitrarily divided is called a MESH. Commercial packages contain programmes for generating appropriate meshes. The choice of an acceptable mesh is a matter of judgement and experience. There is no unique form. There are a number of arcane considerations relating to the more abstract aspects of the theory, but two simple guiding principles are helpful:

  (i) A larger number of smaller elements is needed in regions of high stress concentration.
 (ii) Avoid incorporating long thin elements because they cause computational problems. The analysis may not converge to a solution and the error in the result will be high. It must always be borne in mind that the finite element method produces approximate solutions. Skill and experience is needed to order the analysis so that the error in the solution is a minimum.

With this general background we will now proceed to analyse the reaction of the simple triangular element to applied forces.

6.1 THE CONSTANT STRAIN TRIANGULAR ELEMENT

In this section we shall be concerned with a single triangular element taken from a thin sheet in plane stress. This means that there are no stresses in the direction perpendicular to the plane of the sheet, so the stress system is two-dimensional or BIAXIAL.

In Fig. 6.1 we show the triangular element. We give it the general identity (e) and identify the nodes at the vertices of the triangle by i, j and k in an anti-clockwise order. The element is set up in two-dimensional co-ordinate space with orthogonal x and y axes. Co-ordinates specified relative to these axes are called Global Co-ordinates. In these terms the co-ordinates of the nodes are $(x_i \, y_i) \, (x_j \, y_j)$ and $(x_k \, y_k)$.

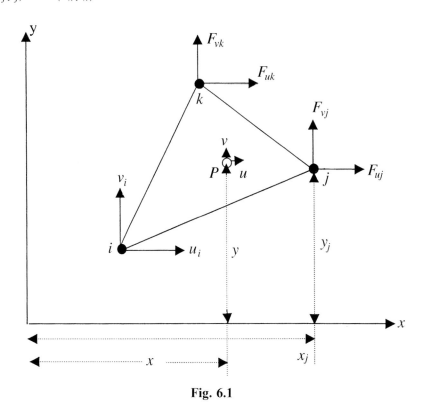

Fig. 6.1

The co-ordinates of a general point P within the element are $(x \, y)$. The nodal displacements are specified by their components in the co-ordinate directions. For example, in Fig. 6.1, u_i and v_i at node i. The six displacement components at the three nodes are called six degrees of freedom and are collected together in the NODAL DISPLACEMENT VECTOR $\{u\}$ where

$$\{u\} = \begin{Bmatrix} u_i \\ v_i \\ u_j \\ v_j \\ u_k \\ v_k \end{Bmatrix} \text{ or alternatively } \begin{Bmatrix} u_i & v_i & u_j & v_j & u_k & v_k \end{Bmatrix}^T$$

to save space. The nodal displacements will distort the element with the result that all points within the element will be displaced. Let the components of the displacement of a point within the element such as P be $(u \, v)$.

Similarly we specify the nodal forces by the two components parallel to the x and y axes. For example at node k (see Fig. 6.1) the components are F_{uk} and F_{vk}. The six nodal force components are collected together in the NODAL FORCE VECTOR $\{F\}$, where

$$\{F\} = \begin{Bmatrix} F_{ui} \\ F_{vi} \\ F_{uj} \\ F_{vj} \\ F_{uk} \\ F_{vk} \end{Bmatrix} \text{ or } \begin{Bmatrix} F_{ui} & F_{vi} & F_{uj} & F_{vj} & F_{uk} & F_{vk} \end{Bmatrix}^T$$

We begin the analysis by making an assumption about the way in which the displacement $(u\ v)$ varies across the element. The higher the order of the terms in the expression for the variation of displacement with x and y (squares cubes and so on) the closer the solution will be to reality. For simplicity, and in order to bring out the essential features of the analysis with the minimum complexity, we choose a linear variation of u and v with x and y. We shall introduce higher order variations in the next chapter.

Our assumption that the displacements within the element, u and v, vary linearly with x and y implies that

$$u = \alpha_1 + \alpha_2 x + \alpha_3 y$$

and

$$v = \alpha_4 + \alpha_5 x + \alpha_6 y \tag{6.1}$$

Where the coefficients α_1 to α_6 are constants for a particular element. If we substitute the nodal co-ordinates in (6.1) then the displacements will be the nodal displacements. This substitution will give six equations, one for each displacement component, which relates the six values of α to the nodal co-ordinates and the nodal displacements viz

$$u_i = \alpha_1 + \alpha_2 x_i + \alpha_3 y_i$$
$$v_i = \alpha_4 + \alpha_5 x_i + \alpha_6 y_i$$
$$u_j = \alpha_1 + \alpha_2 x_j + \alpha_3 y_j \tag{6.2}$$

and similarly for the remaining three nodal displacements. The six equations are gathered together in the following 6×6 matrix.

$$\begin{Bmatrix} u_i \\ v_i \\ u_j \\ v_j \\ u_k \\ v_k \end{Bmatrix} = \begin{bmatrix} 1 & x_i & y_i & 0 & 0 & 0 \\ 0 & 0 & 0 & 1 & x_i & y_i \\ 1 & x_j & y_j & 0 & 0 & 0 \\ 0 & 0 & 0 & 1 & x_j & y_j \\ 1 & x_k & y_k & 0 & 0 & 0 \\ 0 & 0 & 0 & 1 & x_k & y_k \end{bmatrix} \begin{Bmatrix} \alpha_1 \\ \alpha_2 \\ \alpha_3 \\ \alpha_4 \\ \alpha_5 \\ \alpha_6 \end{Bmatrix} \tag{6.3}$$

It might be helpful at this stage to remind ourselves about the relationship between the matrix eqn (6.3) and the separate eqn (6.2) by multiplying out the first row of (6.3) which yields the first

53

eqn in (6.2). This process was explained in section 2.1 of Chapter 2. It gives

$$u_i = 1\alpha_1 + x_i\alpha_2 + y_i\alpha_3 + 0\alpha_4 + 0\alpha_5 + 0\alpha_6$$

or

$$u_i = \alpha_1 + \alpha_2 x_i + \alpha_3 y_i$$

which is the original equation in (6.2). The structure of matrix equation (6.3) is

$$\{u\} = [A]\{\alpha\}$$

If we invert this equation we get

$$\{\alpha\} = [A]^{-1}\{u\} \tag{6.4}$$

This gives the values of the coefficients α in terms of the nodal co-ordinates, which appear in the matrix $[A]^{-1}$, and the nodal displacements, which comprise the nodal displacement vector $\{u\}$. We address the inversion of the matrix $[A]$ later in the chapter and in Appendix (6.1).

6.1.1 THE ELASTIC STRAIN IN THE ELEMENT

If we know the way in which the displacements within the element, u and v, vary with position x and y, we can determine the elastic strains and, from these the stresses. The next step is to establish how this is done.

We relate the displacements to the direct and the shear strains in the following way Mark out a small element of length ∂x in the direction of x (and u) at a point in the finite element.

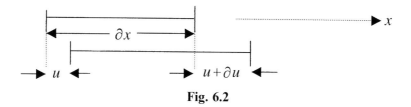

Fig. 6.2

This is the upper line in Fig. 6.2. Let the left hand end of the element be displaced by u and the right hand end by $u + \partial u$. This is represented in the lower line in Fig. 6.2. The length of the element increases because the right hand end is displaced more than the left hand end. Indeed the extension of the element is

$$(u + \partial u) - u = \partial u.$$

The longitudinal strain in the element brought about by these displacements is $\varepsilon_{xx} =$ longitudinal extension/original length or

$$\varepsilon_{xx} = \frac{\partial u}{\partial x} \tag{6.5}$$

The two suffices in the symbol for the strain call for explanation. The first suffix is the direction in which the strain measuring gauge length is oriented, in this case the x direction, and the second suffix is the direction of the displacements at the end of the gauge length, in this case also x. So if the suffices are identical the strain is a direct or tensile strain.

We can get a similar relation to (6.5) by turning Fig. 6.2 through 90° so that it is oriented in the y direction. The element's initial length would be ∂y and the end displacements would be v and $v + \partial v$. Obviously the strain in that direction would be

$$\varepsilon_{yy} = \frac{\partial v}{\partial y} \tag{6.6}$$

Now let us look at shear strains. Here the particle displacement is at right angles to the gauge length. The shear strain on any plane is the angle in radians through which two orthogonal lines in the undistorted body are rotated by the strain. We represent this in Fig. 6.3.

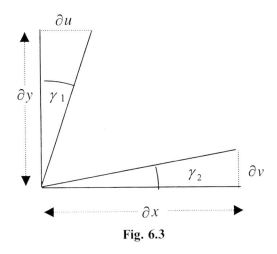

Fig. 6.3

Two lines of length ∂x and ∂y are oriented in the x and y directions respectively and are therefore at right angles. The right hand end of ∂x is displaced in the y direction by ∂v so the line is rotated through an angle

$$\gamma_2 = \frac{\partial v}{\partial x} = \varepsilon_{xy}.$$

Note that the suffices to ε_{xy} mean that the gauge length is in the x-direction and the displacement of the end of the gauge length is in the y-direction. So shear strains have two different suffices. The strain is strictly $\tan \gamma_2$ but since the elastic shear strains that are our concern will be small little error arises from equating the tangent to the angle. Now we displace the upper end of the gauge line ∂y in the x direction by ∂u. This causes a rotation of the line through the angle γ_1 where

$$\gamma_1 = \frac{\partial u}{\partial y} = \varepsilon_{yx}.$$

Here the gauge line is in the y-direction and the end displacement is in the x-direction. The shear strain., which is called the engineering shear strain γ_{xy} is the total rotation of the two lines, that is

$$\gamma_{xy} = \gamma_1 + \gamma_2 = \frac{\partial u}{\partial y} + \frac{\partial v}{\partial x} = \varepsilon_{xy} + \varepsilon_{yx} \tag{6.7}$$

In the case of γ, the suffix xy should be taken to mean that γ is the angular rotation of two initially orthogonal lines that lie in the x–y plane. It is tidy to gather up the three strain expressions (6.5), (6.6) and (6.7) into the strain vector

$$\{\varepsilon\} = \left\{ \begin{array}{c} \varepsilon_{xx} \\ \varepsilon_{yy} \\ \gamma_{xy} \end{array} \right\} = \left\{ \begin{array}{c} \dfrac{\partial u}{\partial x} \\[2mm] \dfrac{\partial v}{\partial y} \\[2mm] \dfrac{\partial u}{\partial y} + \dfrac{\partial v}{\partial x} \end{array} \right\} \tag{6.8}$$

The three equations in (6.8) are used to obtain expressions for the strains by differentiating eqn (6.1).

$$\varepsilon_{xx} = \frac{\partial}{\partial x}(\alpha_1 + \alpha_2 x + \alpha_3 y) = \alpha_2$$

$$\varepsilon_{yy} = \frac{\partial}{\partial y}(\alpha_4 + \alpha_5 x + \alpha_6 y) = \alpha_6 \tag{6.9}$$

$$\gamma_{xy} = \frac{\partial}{\partial y}(\alpha_1 + \alpha_2 x + \alpha_3 y) + \frac{\partial}{\partial x}(\alpha_4 + \alpha_5 x + \alpha_6 y) = (\alpha_3 + \alpha_5).$$

Since all the α coefficients are constant within an element, all the strains are seen from equations (6.9) to be independent of x and y. For this reason the element is referred to as a CONSTANT STRAIN ELEMENT. Because of the constant strain pattern the initially straight edges of the element remain straight in the deformed state. So if a number of such elements are joined together at their corner nodes and remain so joined after straining, then the element edges will still fit together in the deformed state. This is the state of COMPATIBILITY. With this element, compatibility is preserved not only at the nodes but also along the sides of the element. We can now express the strain vector of (6.8) in terms of the α coefficients of (6.9) by recasting (6.9) in matrix form

$$\{\varepsilon\} = \left\{ \begin{array}{c} \varepsilon_{xx} \\ \varepsilon_{yy} \\ \gamma_{xy} \end{array} \right\} = \begin{bmatrix} 0 & 1 & 0 & 0 & 0 & 0 \\ 0 & 0 & 0 & 0 & 0 & 1 \\ 0 & 0 & 1 & 0 & 1 & 0 \end{bmatrix} \left\{ \begin{array}{c} \alpha_1 \\ \alpha_2 \\ \alpha_3 \\ \alpha_4 \\ \alpha_5 \\ \alpha_6 \end{array} \right\} \tag{6.10}$$

It is timely to pause here and gather together the matrix relations we have developed. Equation (6.10) can be represented in abbreviated form as

$$\{\varepsilon\} = [C]\{\alpha\} \tag{6.11}$$

but we have shown that

$$\{\alpha\} = [A]^{-1}\{u\} \qquad [\text{eqn (6.4)}]$$

56

so combining the two relations

$$\{\varepsilon\} = [C][A]^{-1}\{u\} \tag{6.12}$$

if we define the matrix

$$[B] = [C][A]^{-1} \text{ then}$$

$$\{\varepsilon\} = [B]\{u\}. \tag{6.13}$$

The matrix equation is an important one. It relates the strains in the element to the nodal displacements $\{u\}$ through the matrix $[B]$. The derivation of the matrix $[B]$ is a tedious manipulative exercise. Nevertheless we have taken some pains to go through the derivation in Appendix 6.1, in part because it introduces the notion of MATRIX PARTITION which is quite important.

We find that,

$$[B] = \frac{1}{2\Delta} \begin{bmatrix} b_1 & 0 & b_2 & 0 & b_3 & 0 \\ 0 & a_1 & 0 & a_2 & 0 & a_3 \\ a_1 & b_1 & a_2 & b_2 & a_3 & b_3 \end{bmatrix} \tag{6.14}$$

Where Δ is the area of the element and

$$a_1 = (x_k - x_j) \qquad a_2 = (x_i - x_k) \qquad a_3 = (x_j - x_i)$$

$$b_1 = (y_j - y_k) \qquad b_2 = (y_k - y_i) \qquad b_3 = (y_i - y_j) \tag{6.15}$$

It is evident from (6.15) that $[B]$ contains only the nodal co-ordinates, which are known quantities. We have now obtained all the relations we need to deal with elastic strain. The next step is to relate the strains to the stresses.

6.1.2 THE STRESSES IN THE ELEMENT

We now relate the strains to the stresses in the element using the CONSTITUTIVE RELATIONS from the Theory of Elasticity. We develop the relations we need.

For our present purpose we are concerned with a biaxial stress system defined in the following diagram

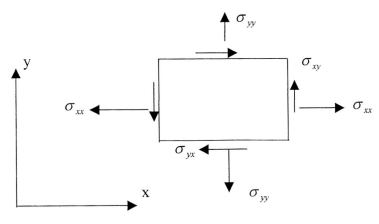

Fig. 6.4

First develop an expression for ε_{xx}, the tensile strain in the direction of x. This has two components, the extension due to σ_{xx} and the contraction due to the Poisson's ratio strain due to σ_{yy}. The following diagram illustrates this:

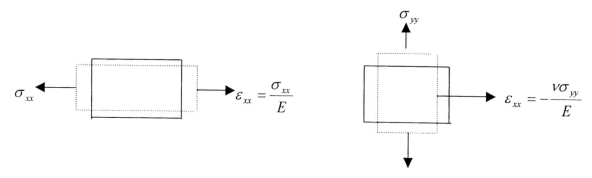

Fig. 6.5

The left-hand diagram shows the extension in the x-direction due to the stress σ_{xx} in that direction. The strain is $\frac{\sigma_{xx}}{E}$. The diagram on the right shows the strain in the x-direction due to the Poisson's ratio contraction arising from the stress σ_{yy} in the y direction. That strain is $-\frac{\nu\sigma_{yy}}{E}$ where ν is Poisson's ratio and E is Young's Modulus. So the total strain in the x-direction arising from the simultaneous application of σ_{xx} and σ_{yy} is

$$\varepsilon_{xx} = \frac{\sigma_{xx}}{E} - \frac{\nu\sigma_{yy}}{E} \tag{6.16}$$

A similar relation with appropriate change of symbols gives the strain in the y-direction

$$\varepsilon_{yy} = \frac{\sigma_{yy}}{E} - \frac{\nu\sigma_{xx}}{E} \tag{6.17}$$

The relation between shear stress and shear strain is shown in the following diagram

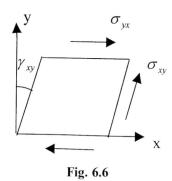

Fig. 6.6

The engineering shear strain γ_{xy} is produced by the shear stress σ_{xy} and the complementary shear stress σ_{yx}. Static equilibrium demands that $\sigma_{xy} = \sigma_{yx}$. The shear modulus G is defined by

$$\gamma_{xy} = \frac{\sigma_{xy}}{G}$$

Elasticity theory gives

$$G = \frac{E}{2(1+\nu)}$$

So

$$\gamma_{xy} = \sigma_{xy}\frac{2(1+\nu)}{E} \tag{6.18}$$

Rearranging eqns (6.16), (6.17) and (6.18) gives

$$\sigma_{xx} = \frac{E}{(1-\nu^2)}(\varepsilon_{xx} + \nu\varepsilon_{yy}) \tag{6.19}$$

$$\sigma_{yy} = \frac{E}{(1-\nu^2)}(\varepsilon_{yy} + \nu\varepsilon_{xx}) \tag{6.20}$$

$$\sigma_{xy} = \frac{E}{2(1+\nu)}\gamma_{xy} = \frac{E}{(1-\nu^2)}\frac{(1-\nu)}{2}\gamma_{xy} \tag{6.21}$$

where we have ordered eqn (6.21) to extract the factor

$$\frac{E}{(1-\nu^2)}$$

which is common to eqns (6.19) and (6.20). Putting eqns (6.19), (6.20) and (6.21) in matrix form

$$\begin{Bmatrix} \sigma_{xx} \\ \sigma_{yy} \\ \sigma_{xy} \end{Bmatrix} = \frac{E}{1-\nu^2}\begin{bmatrix} 1 & \nu & 0 \\ \nu & 1 & 0 \\ 0 & 0 & \frac{(1-\nu)}{2} \end{bmatrix}\begin{Bmatrix} \varepsilon_{xx} \\ \varepsilon_{yy} \\ \gamma_{xy} \end{Bmatrix} \tag{6.22}$$

This matrix equation gives the relation between stress and strain for the biaxial stress state in the constant strain triangular element. It has the condensed form

$$\{\sigma\} = [D]\{\varepsilon\} \tag{6.23}$$

where

$$[D] = \frac{E}{1-\nu^2}\begin{bmatrix} 1 & \nu & 0 \\ \nu & 1 & 0 \\ 0 & 0 & \frac{(1-\nu)}{2} \end{bmatrix} \tag{6.24}$$

is the MATRIX OF ELASTIC CONSTANTS.

The elastic constants matrix (6.24) is for the condition of plane stress. There is a different matrix for plane strain, the condition in which the Poisson's ratio strains in the direction perpendicular to x and y are constrained to be zero. This stress state will not be our concern.

We have established that

$$\{\varepsilon\} = [B]\{u\} \quad [\text{eqn (6.13)}]$$

59

and

$$\{\sigma\} = [\mathbf{D}]\{\varepsilon\} \qquad [\text{eqn (6.23)}]$$

so

$$\{\sigma\} = [\mathbf{D}][\mathbf{B}]\{u\} \tag{6.25}$$

The matrix eqn (6.25) represents an important milestone on the route to the stiffness matrix of the constant strain triangular element. In it we have related the nodal displacements, which are the principal unknowns in the analysis, to the element stresses. We now need to introduce the nodal forces, which will be known quantities, into the analysis. This is achieved by relating the nodal forces to the element stresses. We shall use the Principle of Virtual Work to this end. We could alternatively used the Principle of Stationary Total Potential but we have chosen to give special emphasis to the former.

6.1.3 THE RELATION BETWEEN THE ELEMENT STRESSES AND THE NODAL FORCES

Consider the triangular element of uniform thickness t with nodal forces

$$(F_{ui}, F_{vi}, F_{uj}, F_{vj}, F_{uk}, F_{vk})$$

and nodal displacements

$$(u_i, v_i, u_j, v_j, u_k, v_k)$$

The forces generate stresses in the element. The biaxial stress system is defined in Fig. 6.7.

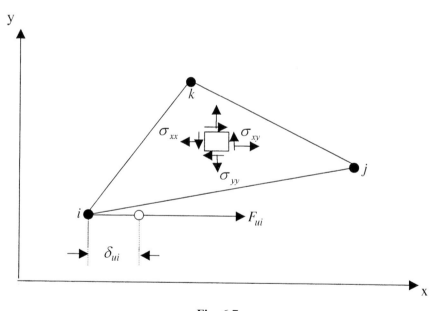

Fig. 6.7

We now keep all the nodal displacements zero except for u_i and then increase u_i by a small virtual displacement δ_{ui}. This is so very small in relation to u_i that the element stresses remain unchanged but the strains will increase by very small increments $\delta\varepsilon_{xx}$, $\delta\varepsilon_{yy}$ and $\delta\gamma_{xy}$. This will produce a very small increase in the elastic strain energy stored in the small section of the element drawn in Fig. 6.7. Our next task is to work out the magnitude of this strain energy increase. This is equal to the work done by the stresses when the faces of an element are displaced by the strain. Consider first the work done by σ_{xx}. Fig. 6.8A shows a small volume element through the thickness t of the finite element and of surface area $dx.dy$. The stress σ_{xx} acts on the face of area $t.dy$.

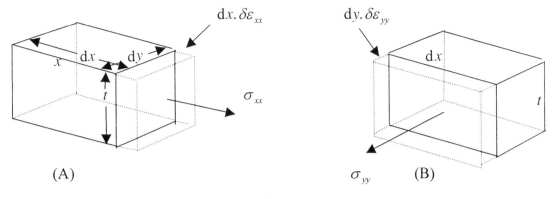

Fig. 6.8

So the force on that face is $\sigma_{xx}.t\,dy$. The displacement of the face due to the strain $\delta\varepsilon_{xx}$ is the product of the strain and the length of the element in the strain direction, that is $dx.\,\delta\varepsilon_{xx}$. The work done is the product of the force and the displacement, because the force is not changed by the virtual displacement, so the work is

$$(\sigma_{xx}.t\,dy).(dx.\delta\varepsilon_{xx}) = \sigma_{xx}.\delta\varepsilon_{xx}(t.dx.dy)$$

but $(t.dx.dy)$ is the volume of the element dV so the work done, which is stored as elastic strain energy is

$$\sigma_{xx}\delta\varepsilon_{xx}.dV \tag{6.26}$$

Similarly, referring to Fig. 6.8B the work done by σ_{yy} during the virtual strain increment $\delta\varepsilon_{yy}$ is

$$(\sigma_{yy}.t\,dx).(dy.\delta\varepsilon_{yy}) = \sigma_{yy}.\delta\varepsilon_{yy}(t.dx.dy) = \sigma_{yy}\delta\varepsilon_{yy}.dV \tag{6.27}$$

The work done by the shear stresses can be worked out from Fig. 6.9.

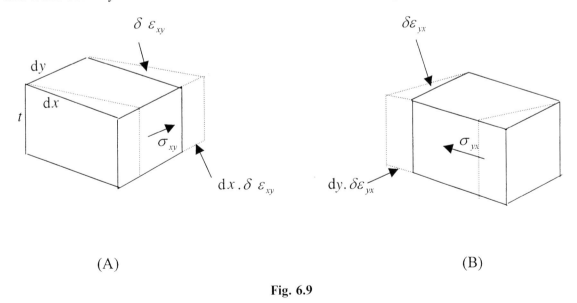

(A)   (B)

Fig. 6.9

61

In Fig. 6.9A in the element $dx.dy.t$, the shear stress σ_{xy} generates a shear force on the end face of $\sigma_{xy}.t.dy$ and the displacement of the face is $dx.\delta\varepsilon_{xy}$. So the work done is

$$(\sigma_{xy}.t.dy).(dx.\delta\varepsilon_{xy}) = \sigma_{xy}.\delta\varepsilon_{xy}.dV \tag{6.28}$$

In Fig. 6.9B, σ_{yx} generates a shear force $\sigma_{yx}.t.dx$ and the end face displaces $dy. \delta\varepsilon_{yx}$. So the work done by σ_{yx} is

$$(\sigma_{yx}.t.dx).(dy.\delta\varepsilon_{yx}) = \sigma_{yx}.\delta\varepsilon_{yx}.dV \tag{6.29}$$

The total work done by the two shear stress components, which contributes to the total elastic strain energy generated by the virtual displacement δu_i, is the sum of (6.28) and (6.29) which is

$$\sigma_{xy}.\delta\varepsilon_{xy}.dV + \sigma_{yx}.\delta\varepsilon_{yx}.dV$$

But $\sigma_{xy} = \sigma_{yx}$ so the shear strain energy is

$$\sigma_{xy}(\delta\varepsilon_{xy} + \delta\varepsilon_{yx})dV \quad \text{and} \quad (\delta\varepsilon_{xy} + \delta\varepsilon_{yx}) = (\delta\gamma_2 + \delta\gamma_1) = \delta\gamma_{xy}$$

so the final expression for the contribution of shear to the stored elastic energy is

$$\sigma_{xy}.\delta\gamma_{xy}.dV \tag{6.30}$$

The total increase of elastic stored energy dU in the element, of volume $dx.dy.t = dV$, due to the virtual displacement δu_i is the sum of (6.26), (6.27) and (6.30), which is

$$dU = (\sigma_{xx}\delta\varepsilon_{xx} + \sigma_{yy}\delta\varepsilon_{yy} + \sigma_{xy}\delta\gamma_{xy})dV \tag{6.31}$$

The elastic strain energy increase in the whole triangular element ΔU is obtained by summing (that is integrating) this expression over the total volume. That is

$$\Delta U = \int_V (\sigma_{xx}\delta\varepsilon_{xx} + \sigma_{yy}\delta\varepsilon_{yy} + \sigma_{xy}\delta\gamma_{xy})dV \tag{6.32}$$

Now throughout the whole volume V of the triangular element both the stresses and the strains are constant so the terms inside the ( ) brackets are constant. So we can take these terms outside the integral sign and the integral becomes very simple. It is

$$\Delta U = (\sigma_{xx}\delta\varepsilon_{xx} + \sigma_{yy}\delta\varepsilon_{yy} + \sigma_{xy}\delta\gamma_{xy}) \int_V dV$$

$$= (\sigma_{xx}\delta\varepsilon_{xx} + \sigma_{yy}\delta\varepsilon_{yy} + \sigma_{xy}\delta\gamma_{xy})V. \tag{6.33}$$

In the next chapter, where we analyse more complex elements, we shall face situations in which the stresses vary across the element and the integral in (6.32) requires special methods for its evaluation.

For our future needs it will be helpful to express (6.33) in matrix form, that is

$$\Delta U = V\{ \delta\varepsilon_{xx} \quad \delta\varepsilon_{yy} \quad \delta\gamma_{xy} \} \begin{Bmatrix} \sigma_{xx} \\ \sigma_{yy} \\ \sigma_{xy} \end{Bmatrix} \tag{6.34}$$

Note that when we multiply out the above matrix equation we get

$$\Delta U = V(\delta\varepsilon_{xx}\sigma_{xx} + \delta\varepsilon_{yy}\sigma_{yy} + \delta\gamma_{xy}\sigma_{xy})$$

which is identical to (6.33) because reversing the order of the product terms in the ( ) brackets does not change the value of the product.

It is convenient to express the matrix equation (6.34) in shorthand form

$$\Delta U = V\{\delta\varepsilon\}^T\{\sigma\} \tag{6.35}$$

We derived a relation between the nodal displacement vector and the vector of elastic strains in the triangular element

$$\varepsilon = [\mathbf{B}]\{u\} \qquad [\text{eqn (6.13)}]$$

A similar relation must exist between the virtual nodal displacements $\{\delta u\}$ and the virtual strains produced $\{\delta\varepsilon\}$ so

$$\{\delta\varepsilon\} = [\mathbf{B}]\{\delta u\} \tag{6.36}$$

Note that in equation (6.13) $\{\varepsilon\}$ is a column vector that is

$$\begin{Bmatrix} \varepsilon_{xx} \\ \varepsilon_{yy} \\ \gamma_{xy} \end{Bmatrix}$$

therefore in (6.36) $\{\delta\varepsilon\}$ must also be a column vector. However in the matrix equation (6.35) we require the virtual strains in a row vector. To achieve this we transpose the column vector $\{\delta\varepsilon\}$ into the row vector $\{\delta\varepsilon\}^T$. This explains the form of (6.35). We shall need to express the virtual strains in (6.35) in terms of the virtual nodal displacements. For this purpose we use equation (6.36) but we need to transpose $\{\delta\varepsilon\}$ in (6.36) into $\{\delta\varepsilon\}^T$ which appears in (6.35). Now $\{\delta\varepsilon\}^T$ therefore is $([\mathbf{B}]\{\delta u\})^T$ so we are concerned with the transpose of the product of two matrices.

In Appendix 6.2, we show by example that the transpose of the product of two matrices is the product of the two transposed matrices with their order reversed. So if

$$[\mathbf{L}] = [\mathbf{M}][\mathbf{N}] \quad [\mathbf{L}]^T = [\mathbf{N}]^T[\mathbf{M}]^T$$

similarly if

$$\{\delta\varepsilon\} = [\mathbf{B}]\{\delta u\} \quad \{\delta\varepsilon\}^T = \{\delta u\}^T[\mathbf{B}]^T \tag{6.37}$$

It will be helpful to present this matrix equation in full

$$\{\delta\varepsilon\}^T = \{\, \delta\varepsilon_{xx} \quad \delta\varepsilon_{yy} \quad \delta\gamma_{xy} \,\} = \{\, \delta u_i \quad \delta v_i \quad \delta u_j \quad \delta v_j \quad \delta u_k \quad \delta v_k \,\} \begin{bmatrix} b_1 & 0 & a_1 \\ 0 & a_1 & b_1 \\ b_2 & 0 & a_2 \\ 0 & a_2 & b_2 \\ b_3 & 0 & a_3 \\ 0 & a_3 & b_3 \end{bmatrix} \tag{6.38}$$

The a and b terms in the matrix were defined in equation (6.15).

63

We now return to Fig. 6.7 where we gave node i in the triangular element a virtual displacement δu_i with all other degrees of freedom fixed. We have calculated the increase of elastic stored energy. The balancing decrease of potential energy of the nodal applied force is

$$F_{ui}\delta u_i.$$

We can now complete the statement of the Virtual Work Principle that the increase of internal strain energy in a virtual displacement = the fall of potential energy of the applied forces, so

$$V(\sigma_{xx}\delta\varepsilon_{xx} + \sigma_{yy}\delta\varepsilon_{yy} + \sigma_{xy}\delta\gamma_{xy}) = F_{ui}\delta u_i$$

The compact matrix form of the left hand side of this equation was shown to be

$$V\{\delta\varepsilon\}^T\{\sigma\} \quad \text{in eqn (6.35) so}$$
$$V\{\delta\varepsilon\}^T\{\sigma\} = F_{ui}\,\delta u_i \tag{6.39}$$

We have shown in (6.37) that

$$\{\delta\varepsilon\}^T = \{\delta u\}^T[\mathbf{B}]^T$$
$$= \{\,\delta u_i \quad \delta v_i \quad \delta u_j \quad \delta v_j \quad \delta u_k \quad \delta v_k\,\}[\mathbf{B}]^T \tag{6.40}$$

But, since we displace only u_i and fix all the other degrees of freedom, only δu_i is non-zero. So if we substitute this in (6.39) and expand $[\mathbf{B}]^T$ we get

$$\{\delta\varepsilon\}^T = \{\,\delta u_i \quad 0 \quad 0 \quad 0 \quad 0 \quad 0\,\}\begin{bmatrix} b_1 & 0 & a_1 \\ 0 & a_1 & b_1 \\ b_2 & 0 & a_2 \\ 0 & a_2 & b_2 \\ b_3 & 0 & a_3 \\ 0 & a_3 & b_3 \end{bmatrix}$$

Multiplying out the right-hand side

$$\{\delta\varepsilon\}^T = \delta u_i \, [b_1 \quad 0 \quad a_1]$$

The work balance of eqn (6.39) becomes

$$V.\delta u_i \, [b_1 \quad 0 \quad a_1] \, \{\sigma\} = F_{ui}\,\delta u_i$$

And $\{\delta u_i\}$ cancels so

$$V[b_1 \quad 0 \quad a_1]\{\sigma\} = F_{ui} \tag{6.41}$$

At this point in this marathon analysis we can draw breath and note with some comfort that we have now related the nodal force F_{ui} to the stresses produced by it $\{\sigma\}$ which was the objective we set ourselves at the outset of this section.

Equation (6.41) gives the stresses generated by F_{ui}. We now need to derive expressions for the stresses generated by the other five load components. To this end we now displace node i by a virtual displacement δv_i with all other degrees of freedom fixed to zero. Correspondingly only nodal force F_{vi} suffers a displacement.

$$\{\delta\varepsilon\}^T = \{0 \quad \delta v_i \quad 0 \quad 0 \quad 0 \quad 0\} \begin{bmatrix} b_1 & 0 & a_1 \\ 0 & a_1 & b_1 \\ b_2 & 0 & a_2 \\ 0 & a_2 & b_2 \\ b_3 & 0 & a_3 \\ 0 & a_3 & b_3 \end{bmatrix}$$

so

$$\{\delta\varepsilon\}^T = \delta v_i [0 \quad a_1 \quad b_1]$$

and the work balance from equation (6.35) is

$$V \delta v_i [0 \quad a_1 \quad b_1]\{\sigma\} = F_{vi}\, \delta v_i$$

and δv_i cancels and

$$V [0 \quad a_1 \quad b_1]\{\sigma\} = F_{vi} \tag{6.42}$$

We follow the same procedure for each of the other four degrees of freedom in turn displacing u_j, v_j, u_k and v_k keeping all but the operative degree of freedom zero. This gives us relations for the other four nodal forces. They are

$$V [b_2 \quad 0 \quad a_2]\{\sigma\} = F_{uj} \tag{6.43}$$

$$V [0 \quad a_2 \quad b_2]\{\sigma\} = F_{vj} \tag{6.44}$$

$$V [b_3 \quad 0 \quad a_3]\{\sigma\} = F_{uk} \tag{6.45}$$

$$V [0 \quad a_3 \quad b_3]\{\sigma\} = F_{vk} \tag{6.46}$$

We gather together the six equations (6.41) to (6.46) into a single matrix equation

$$V \begin{bmatrix} b_1 & 0 & a_1 \\ 0 & a_1 & b_1 \\ b_2 & 0 & a_2 \\ 0 & a_2 & b_2 \\ b_3 & 0 & a_3 \\ 0 & a_3 & b_3 \end{bmatrix} \{\sigma\} = \begin{Bmatrix} F_{ui} \\ F_{vi} \\ F_{uj} \\ F_{vj} \\ F_{uk} \\ F_{vk} \end{Bmatrix} \tag{6.47}$$

This is the important relation we seek between the (constant) element stresses and the element nodal force vector. Equation (6.47) has the condensed structure

$$V[\mathbf{B}]^T\{\sigma\} = \{\mathbf{F}\} \tag{6.48}$$

65

We have established that

$$\{\sigma\} = [D]\,[B]\{u\} \qquad [\text{eqn } (6.25)]$$

substituting this in (6.48)

$$V[B]^T[D]\,[B]\{u\} = \{F\} \tag{6.49}$$

This is one of the most important relations in finite element theory. It relates nodal forces and nodal displacements. It has the general form

$$[K]\{u\} = \{F\} \tag{6.50}$$

where if we compare (6.49) and (6.50) we see that

$$[K] = V[B]^T[D]\,[B] \tag{6.51}$$

and [K] is called the STIFFNESS MATRIX.

For the constant strain triangular element we have all the information to evaluate the three matrices that make up the stiffness matrix. So [K] can be evaluated and we know the applied nodal forces {F} so the only unknowns in (6.50) are the nodal displacements. We set up the stiffness matrices (6.50) for all the elements and the assemble them in exactly the same way that we did for the linear array of springs. This gives us the global stiffness matrix. Inverting this matrix gives {u} for the whole element mesh and we have achieved a solution.

The element stiffness matrix is so important that it is worth looking at it in more detail. Expand (6.51) by writing out the constituent matrices

$$[K] = V\frac{1}{2\Delta}\frac{E}{1-\nu^2}
\begin{bmatrix}
b_1 & 0 & a_1 \\
0 & a_1 & b_1 \\
b_2 & 0 & a_2 \\
0 & a_2 & b_2 \\
b_3 & 0 & a_3 \\
0 & a_3 & b_3
\end{bmatrix}
\begin{bmatrix}
1 & \nu & 0 \\
\nu & 1 & 0 \\
0 & 0 & \frac{1-\nu}{2}
\end{bmatrix}
\begin{bmatrix}
b_1 & 0 & b_2 & 0 & b_3 & 0 \\
0 & a_1 & 0 & a_2 & 0 & a_3 \\
a_1 & b_1 & a_2 & b_2 & a_3 & b_3
\end{bmatrix} \tag{6.52}$$

We have gathered all the matrix multiplying constants together at the beginning of the expression. The order of the three matrices in (6.52) which are multiplied together is

$$[6 \times 3]\,[3 \times 3]\,[3 \times 6\}. \tag{6.53}$$

Recall that we can multiply matrices together only if the number of columns in the first is equal to the number of rows in the second. This condition is satisfied by (6.53). The product of the second and third matrices has order $[3 \times 6]$ and the product of the first matrix and this has order $[6 \times 6]$. So the stiffness matrix, which is (6.52), is a $[6 \times 6]$ matrix. The stiffness matrix equation (6.50)

$$[K]\{u\} = \{F\}$$

therefore has the structure

$$
\begin{bmatrix}
k_{11} & k_{12} & k_{13} & k_{14} & k_{15} & k_{16} \\
k_{21} & k_{22} & k_{23} & k_{24} & k_{25} & k_{26} \\
k_{31} & k_{32} & k_{33} & k_{34} & k_{35} & k_{36} \\
k_{41} & k_{42} & k_{43} & k_{44} & k_{45} & k_{46} \\
k_{51} & k_{52} & k_{53} & k_{54} & k_{55} & k_{56} \\
k_{61} & k_{62} & k_{63} & k_{64} & k_{65} & k_{66}
\end{bmatrix}
\begin{Bmatrix}
u_i \\ v_i \\ u_j \\ v_j \\ u_k \\ v_k
\end{Bmatrix}
=
\begin{Bmatrix}
F_{ui} \\ F_{vi} \\ F_{uj} \\ F_{vj} \\ F_{uk} \\ F_{vk}
\end{Bmatrix}
\tag{6.54}
$$

where we have laid out the 36 stiffness coefficients of the stiffness matrix of the element.

This is a good point in the discussion to use (6.54) to reinforce the section in Chapter 3 about the significance of the terms in the stiffness matrix. Multiplying out the third row of (6.54) by way of example gives the following equation for F_{uj}

$$k_{31}u_i + k_{32}v_i + k_{33}u_j + k_{34}v_j + k_{35}u_k + k_{36}v_k = F_{uj}$$

The terms on the left-hand side of the equation are the six contributions to the force F_{uj} due to displacements in turn of the degrees of freedom u_i to v_k under conditions where all displacements other than the operative one are fixed at zero. So, for example, $k_{32}\,v_i$ is the force at node j in the direction of u (F_{uj}) due to a displacement of node i of v_i with degrees of freedom $u_i\,u_j\,v_j\,u_k$ and v_k constrained at zero during the displacement v_i. Similarly $k_{35}\,u_k$ is the force at node j in the u direction due to a displacement of node k, u_k, with degrees of freedom u_i, v_i, u_j, v_j and v_k constrained to be zero.

It is also worth emphasising that there are no zeros in the element stiffness matrix. This is because in the element all nodes are nearest neighbour to all the other nodes so every degree of freedom contributes to all the force components. Zeros appear when the stiffness matrices of several elements are assembled in a global matrix. Then there are nodes in one element which are so far separated from nodes in another element that fixed nodes between them prevent force transmission between them.

This completes the development of the analysis needed to solve a finite element model using the constant strain triangular element. We now demonstrate how the analysis works out in practice by working through an absurdly simple example with only two elements. This will not produce a useful stress distribution. But the simplicity of the model will make it feasible to work through all the manipulations in revealed detail so all the steps are completely transparent. After this we shall present a solution obtained by using a commercial finite element package with a sufficient number of elements to produce an output that makes engineering sense.

APPENDIX 6.1   INVERSION OF THE MATRIX [A] AND DERIVATION OF THE STRAIN MATRIX [B]

We begin by reproducing equation (6.3)

$$
\begin{Bmatrix}
u_i \\ v_i \\ u_j \\ v_j \\ u_k \\ v_k
\end{Bmatrix}
=
\begin{bmatrix}
1 & x_i & y_i & 0 & 0 & 0 \\
0 & 0 & 0 & 1 & x_i & y_i \\
1 & x_j & y_j & 0 & 0 & 0 \\
0 & 0 & 0 & 1 & x_j & y_j \\
1 & x_k & y_k & 0 & 0 & 0 \\
0 & 0 & 0 & 1 & x_k & y_k
\end{bmatrix}
\begin{Bmatrix}
\alpha_1 \\ \alpha_2 \\ \alpha_3 \\ \alpha_4 \\ \alpha_5 \\ \alpha_6
\end{Bmatrix}
\tag{6.1.1}
$$

or in condensed form

$$\{u\} = [A]\{\alpha\}$$

We require $\{\alpha\}$ in terms of $\{u\}$ that is

$$\{\alpha\} = [A]^{-1}\{u\}$$

Inverting [A] directly is over-facing so we first re-order the [A] matrix (6.1.1) to simplify the inversion. Equation (6.1.1) represents six equations. We can legitimately list these equations in any order, accordingly we re-order them as follows

$$\begin{Bmatrix} u_i \\ u_j \\ u_k \\ v_i \\ v_j \\ v_k \end{Bmatrix} = \begin{bmatrix} 1 & x_i & y_i & 0 & 0 & 0 \\ 1 & x_j & y_j & 0 & 0 & 0 \\ 1 & x_k & y_k & 0 & 0 & 0 \\ 0 & 0 & 0 & 1 & x_i & y_i \\ 0 & 0 & 0 & 1 & x_j & y_j \\ 0 & 0 & 0 & 1 & x_k & y_k \end{bmatrix} \begin{Bmatrix} \alpha_1 \\ \alpha_2 \\ \alpha_3 \\ \alpha_4 \\ \alpha_5 \\ \alpha_6 \end{Bmatrix} \qquad (6.1.2)$$

We can satisfy ourselves that this re-ordering is valid by multiplying out the same equation, for example that for u_k, in both (6.1.1 and 6.1.2) and showing that they are identical. The equation derives from row 5 in (6.1.1) and from row 3 in (6.1.2).

The next step is to PARTITION or split up this matrix into four 3×3 matrices, two of which contain terms that are all zero an therefore contribute no terms to the equations, so we ignore them. We separate out the two non-zero 3×3 matrices which are

$$\begin{Bmatrix} u_i \\ u_j \\ u_k \end{Bmatrix} = \begin{bmatrix} 1 & x_i & y_i \\ 1 & x_j & y_j \\ 1 & x_k & y_k \end{bmatrix} \begin{Bmatrix} \alpha_1 \\ \alpha_2 \\ \alpha_3 \end{Bmatrix} \qquad (6.1.3)$$

and

$$\begin{Bmatrix} v_i \\ v_j \\ v_k \end{Bmatrix} = \begin{bmatrix} 1 & x_i & y_i \\ 1 & x_j & y_j \\ 1 & x_k & y_k \end{bmatrix} \begin{Bmatrix} \alpha_4 \\ \alpha_5 \\ \alpha_6 \end{Bmatrix} \qquad (6.1.4)$$

We can now, without too much hazard, invert these two 3×3 matrices and then combine them into the final 6x6 matrix. We note with some comfort that the two matrices are identical.

To invert the first matrix (6.1.3), we first evaluate the co-factors with appropriate sign

$$\begin{bmatrix} (x_j y_k - x_k y_j) & -(y_k - y_j) & (x_k - x_j) \\ -(x_i y_k - x_k y_i) & (y_k - y_i) & -(x_k - x_i) \\ (x_i y_j - x_j y_i) & -(y_j - y_i) & (x_j - x_i) \end{bmatrix}$$

Transpose the co-factor matrix and absorb the minus signs.

$$\begin{bmatrix} (x_j y_k - x_k y_j) & (x_k y_i - x_i y_k) & (x_i y_j - x_j y_i) \\ (y_j - y_k) & (y_k - y_i) & (y_i - y_j) \\ (x_k - x_j) & (x_i - x_k) & (x_j - x_i) \end{bmatrix}$$

Divide by the determinant of the original matrix (6.1.3) which is

$$1(x_j y_k - x_k y_j) - x_i(y_k - y_j) + y_i(x_k - x_j)$$

this turns out to be exactly twice the area of the element Δ so the inverted matrix equation is

$$\begin{Bmatrix} \alpha_1 \\ \alpha_2 \\ \alpha_3 \end{Bmatrix} = \frac{1}{2\Delta} \begin{bmatrix} (x_j y_k - x_k y_j) & (x_k y_i - x_i y_k) & (x_i y_j - x_j y_i) \\ (y_j - y_k) & (y_k - y_i) & (y_i - y_j) \\ (x_k - x_j) & (x_i - x_k) & (x_j - x_i) \end{bmatrix} \begin{Bmatrix} u_i \\ u_j \\ u \end{Bmatrix} \tag{6.1.5}$$

Since the matrix in the second equation (6.1.4) is identical then so will its inverse so the inverted equation (6.1.4) is

$$\begin{Bmatrix} \alpha_4 \\ \alpha_5 \\ \alpha_6 \end{Bmatrix} = \frac{1}{2\Delta} \begin{bmatrix} (x_j y_k - x_k y_j) & (x_k y_i - x_i y_k) & (x_i y_j - x_j y_i) \\ (y_j - y_k) & (y_k - y_i) & (y_i - y_j) \\ (x_k - x_j) & (x_i - x_k) & (x_j - x_i) \end{bmatrix} \begin{Bmatrix} v_i \\ v_j \\ v_k \end{Bmatrix} \tag{6.1.6}$$

Combine these two matrix equations into a single 6x6 matrix by putting the terms in the appropriate rows and columns

$$\begin{Bmatrix} \alpha_1 \\ \alpha_2 \\ \alpha_3 \\ \alpha_4 \\ \alpha_5 \\ \alpha_6 \end{Bmatrix} = \frac{1}{2\Delta} \begin{bmatrix} (x_j y_k - x_k y_j) & 0 & (x_k y_i - x_i y_k) & 0 & (x_i y_j - x_j y_i) & 0 \\ (y_j - y_k) & 0 & (y_k - y_i) & 0 & (y_i - y_j) & 0 \\ (x_k - x_j) & 0 & (x_i - x_k) & 0 & (x_j - x_i) & 0 \\ 0 & (x_j y_k - x_k y_j) & 0 & (x_k y_i - x_i y_k) & 0 & (x_i y_j - x_j y_i) \\ 0 & (y_j - y_k) & 0 & (y_k - y_i) & 0 & (y_i - y_j) \\ 0 & (x_k - x_j) & 0 & (x_i - x_k) & 0 & (x_j - x_i) \end{bmatrix} \begin{Bmatrix} u_i \\ v_i \\ u_j \\ v_j \\ u_k \\ v_k \end{Bmatrix}$$

This achieves the inversion of the original matrix equation, that is

$$\{\alpha\} = [A]^{-1}\{u\}$$

The Elastic Strain Matrix [B]

The matrix [B] relates the elastic strain vector $\{\varepsilon\}$ and the nodal displacement vector $\{u\}$

$$\{\varepsilon\} = [B]\{u\}$$

We have seen in equations (6.12) and (6.13) that

$$[B] = [C][A]^{-1} \quad \text{and so} \quad \{\varepsilon\} = [C][A]^{-1}\{u\}.$$

Substituting explicitly for [C] from equation (6.10) and expanding $\{\varepsilon\}$

$$\begin{Bmatrix} \varepsilon_{xx} \\ \varepsilon_{yy} \\ \varepsilon_{xy} \end{Bmatrix} = \begin{bmatrix} 0 & 1 & 0 & 0 & 0 & 0 \\ 0 & 0 & 0 & 0 & 0 & 1 \\ 0 & 0 & 1 & 0 & 1 & 0 \end{bmatrix} [A]^{-1}\{u\} \tag{6.1.7}$$

69

BASIC PRINCIPLES OF THE FINITE ELEMENT METHOD

The matrix product $[C][A]^{-1}$ can readily be evaluated because of the large proportion of zeros in [C]. The product of the 3×6 and the 6×6 matrices is a 3×6 matrix which is [B]. The first row of $\{B\}$ is obtained by multiplying $[A]^{-1}$ by the first row of [C]. This has only one non-zero term – the second one – which is unity. So the first row of [B] is identical to the second row of $[A]^{-1}$. By similar a argument, the second row of [B] is identical to the 6th row of $[A]^{-1}$ and the 3rd row of [B] is the sum of the 3rd and 5th rows of $[A]^{-1}$. The final form of (6.1.7) therefore is

$$\begin{Bmatrix} \varepsilon_{xx} \\ \varepsilon_{yy} \\ \varepsilon_{xy} \end{Bmatrix} = \frac{1}{2\Delta} \begin{bmatrix} (y_j - y_k) & 0 & (y_k - y_i) & 0 & (y_i - y_j) & 0 \\ 0 & (x_k - x_j) & 0 & (x_i - x_k) & 0 & (x_j - x_i) \\ (x_k - x_j) & (y_j - y_k) & (x_i - x_k) & (y_k - y_i) & (x_j - x_i) & (y_i - y_j) \end{bmatrix} \begin{Bmatrix} u_i \\ v_i \\ u_j \\ v_j \\ u_k \\ v_k \end{Bmatrix}$$

We substitute the abbreviated form of the [B] matrix from equation 6.14 and

$$\begin{Bmatrix} \varepsilon_{xx} \\ \varepsilon_{yy} \\ \varepsilon_{xy} \end{Bmatrix} = \frac{1}{2\Delta} \begin{bmatrix} b_1 & 0 & b_2 & 0 & b_3 & 0 \\ 0 & a_1 & 0 & a_2 & 0 & a_3 \\ a_1 & b_1 & a_2 & b_2 & a_3 & b_3 \end{bmatrix} \begin{Bmatrix} u_i \\ v_i \\ u_j \\ v_j \\ u_k \\ v_k \end{Bmatrix}$$

where

$$a_1 = (x_k - x_j) \qquad a_2 = (x_i - x_k) \qquad a_3 = (x_j - x_i)$$
$$b_1 = (y_j - y_k) \qquad b_2 = (y_k - y_i) \qquad b_3 = (y_i - y_j).$$

APPENDIX 6.2 PROOF BY EXAMPLE OF THE RELATION FOR THE TRANSPOSE OF THE PRODUCT OF TWO MATRICES

Let [L] be the product of two matrices [M] and [N], that is

$$[L] = [M][N]$$

We wish to show that

$$[L]^T = [N]^T[M]^T$$

$$\text{let } [M] = \begin{bmatrix} 1 & 2 & 1 \\ 3 & 4 & 2 \\ 3 & 1 & 1 \end{bmatrix} \quad \text{and} \quad [N] = \begin{bmatrix} 1 & 0 \\ 0 & 3 \\ 2 & 1 \end{bmatrix}$$

70

then

$$[L] = [M][N] = \begin{bmatrix} 1 & 2 & 1 \\ 3 & 4 & 2 \\ 3 & 1 & 1 \end{bmatrix} \begin{bmatrix} 1 & 0 \\ 0 & 3 \\ 2 & 1 \end{bmatrix} = \begin{bmatrix} 3 & 7 \\ 7 & 14 \\ 5 & 4 \end{bmatrix}$$

following the rules for matrix multiplication in Section 2.2.

$$\text{so} \quad [L]^T = \begin{bmatrix} 3 & 7 & 5 \\ 7 & 14 & 4 \end{bmatrix}$$

$$\text{now} \quad [N]^T[M]^T = \begin{bmatrix} 1 & 0 \\ 0 & 3 \\ 2 & 1 \end{bmatrix}^T \begin{bmatrix} 1 & 2 & 1 \\ 3 & 4 & 2 \\ 3 & 1 & 1 \end{bmatrix}^T = \begin{bmatrix} 1 & 0 & 2 \\ 0 & 3 & 1 \end{bmatrix} \begin{bmatrix} 1 & 3 & 3 \\ 2 & 4 & 1 \\ 1 & 2 & 1 \end{bmatrix} = \begin{bmatrix} 3 & 7 & 5 \\ 7 & 14 & 5 \end{bmatrix}$$

which is the same as $[L]^T$ and confirms the proposition

6.2 AN EXAMPLE WHICH USES TRIANGULAR CONSTANT STRAIN ELEMENTS

In this example we make a very crude estimate of the stress distribution in a loaded plate that is fixed at one edge. We use the triangular constant strain element for which we have developed the stiffness matrix in section 6.1. The purpose of the example is to show the detailed working that yields the stiffness matrices of the elements and then assembles them to give the global stiffness matrix of the whole plate. To keep the arithmetic within bounds we use only two elements. This gives a very crude estimate of the stress distribution but it illustrates the details of the analysis without too much numerical indigestion and makes it feasible to display all the steps in the analysis in complete detail.

   At the end ot the section the result is quoted of a calculation using the LUSAS package with a meaningful number of elements which gives a more useful solution.

   The geometry of the plate and the location and magnitude of the applied loads are defined in Fig. 6.10. To contribute an element of variety we use imperial units.

   The left-hand edge of the plate is fixed
   Young's modulus of the plate material is 30×10^6 lbf in^{-2}
   Poisson's ratio for the plate material is 1/3.
   The plate thickness is uniform and is 0.1 in.

   We divide the plate into two elements. These are shown in Fig. 6.11 as also are the element numbers, in brackets, and the node numbers.
   Let us remind ourselves of the procedure we shall follow in the analysis.

1. We determine the stiffness matrices of the two elements 1 and 2, which we shall call $[K_1]$ and $[K_2]$.
2. We recall that $[K] = t.\Delta. [B]^T [D] [B]$

71

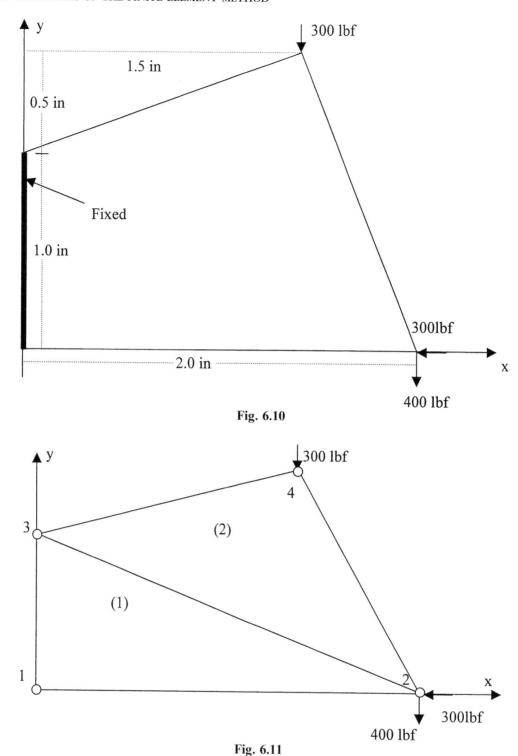

Fig. 6.10

Fig. 6.11

where t is the plate thickness, Δ is the area of the element, and [B] is the strain matrix in

$$\{\varepsilon\} = [B]\{u\}.$$

$\{\varepsilon\}$ is the uniform strain in the element and $\{u\}$ is the vector of nodal displacements. [B] comprises a matrix in which the terms are differences of nodal co-ordinates. We call the strain matrices for elements (1) and (2) $[B_1]$ and $[B_2]$ respectively. The elasticity matrix [D] is the same for both elements, as are t and Δ. So the two stiffness matrices for the elements are

$$[K_1] = t.\Delta.[B_1]^T[D][B_1]$$

$$[K_2] = t.\Delta.[B_2]^T[D][B_2]$$

3. We assemble the two matrices $[K_1]$ and $[K_2]$, which will each be of order $[6 \times 6]$, into a single Global Matrix. This will be of order $[8 \times 8]$ because there are 4 nodes in total each with two displacement components and with two force components making 8 degrees of freedom for the model.
4. We then condense the Global Matrix by eliminating the rows and columns corresponding to the fixed displacement components and the unknown force components. This gives a $[4 \times 4]$ matrix.
5. The condensed matrix is inverted to give the four unknown displacements.
6. These nodal displacements, together with the zero displacements of the fixed nodes, are used to calculate the element stresses using

For element (1) $\{\sigma\} = [D][B_1]\{u\}$
For element (2) $\{\sigma\} = [D][B_2]\{u\}$.

The detailed analysis follows

6.2.1   THE STIFFNESS MATRIX FOR ELEMENT (1)

We draw the element and its applied loads in Fig. 6.12. Note that we can assign the two loads at node 2 to element (1) or to element (2) or one to each. In this case we have chosen to assign both loads to element (1). The node numbering sequence is anti-clockwise.

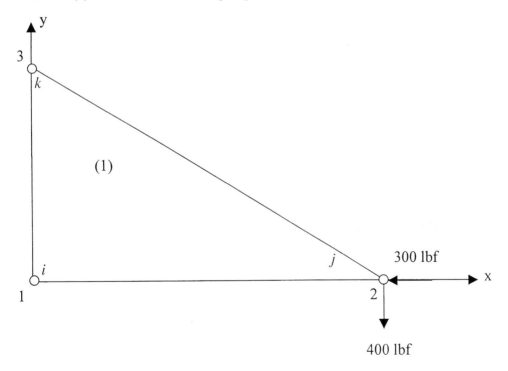

Element (1)

Fig. 6.12

73

BASIC PRINCIPLES OF THE FINITE ELEMENT METHOD

The following table assembles all the data for the element.

Node	x	y	u	v	F_u	F_v
i (1)	0	0	0	0	F_{u1}	F_{v1}
j (2)	2	0	u_2	v_2	-300	-400
k (3)	0	1	0	0	F_{u3}	F_{v3}

The element area is 1.0 in^2
The elasticity matrix is, in units of lbf in^{-2},

$$[D] = \frac{E}{(1-\nu^2)} \cdot \begin{bmatrix} 1 & \nu & 0 \\ \nu & 1 & 0 \\ 0 & 0 & \frac{1-\nu}{2} \end{bmatrix} = 33.75 \times 10^6 \begin{bmatrix} 1 & 1/3 & 0 \\ 1/3 & 1 & 0 \\ 0 & 0 & 1/3 \end{bmatrix}$$

The matrix relation between the elastic strains in the element and the nodal displacements is

$$\{\varepsilon\} = [B]\{u\}$$

In Section 6.1.1 we derived the expression for [B] which is given in eqns (6.14) and (6.15). We reproduce it here

$$[B] = \frac{1}{2\Delta} \begin{bmatrix} (y_j - y_k) & 0 & (y_k - y_i) & 0 & (y_i - y_j) & 0 \\ 0 & (x_k - x_j) & 0 & (x_i - x_k) & 0 & (x_j - x_i) \\ (x_k - x_j) & (y_j - y_k) & (x_i - x_k) & (y_k - y_i) & (x_j - x_i) & (y_i - y_j) \end{bmatrix}$$

Δ is the area of the element.
We substitute the nodal co-ordinates of element (1) in [B] and get

$$[B_1] = \frac{1}{2x1} \begin{bmatrix} -1 & 0 & 1 & 0 & 0 & 0 \\ 0 & -2 & 0 & 0 & 0 & 2 \\ -2 & -1 & 0 & 1 & 2 & 0 \end{bmatrix}$$

so $[B_1]^T$ the transpose of $[B_1]$ is obtained by interchanging the rows and columns viz:

$$[B_1]^T = \frac{1}{2} \begin{bmatrix} -1 & 0 & -2 \\ 0 & -2 & -1 \\ 1 & 0 & 0 \\ 0 & 0 & 1 \\ 0 & 0 & 2 \\ 0 & 2 & 0 \end{bmatrix}$$

74

and the matrix product

$$[B_1]^T [D] [B_1]$$

is

$$\frac{33.75 \times 10^6}{4} \begin{bmatrix} -1 & 0 & -2 \\ 0 & -2 & -1 \\ 1 & 0 & 0 \\ 0 & 0 & 1 \\ 0 & 0 & 2 \\ 0 & 2 & 0 \end{bmatrix} \begin{bmatrix} 1 & \frac{1}{3} & 0 \\ \frac{1}{3} & 1 & 0 \\ 0 & 0 & \frac{1}{3} \end{bmatrix} \begin{bmatrix} -1 & 0 & 1 & 0 & 0 & 0 \\ 0 & -2 & 0 & 0 & 0 & 2 \\ -2 & -1 & 0 & 1 & 2 & 0 \end{bmatrix}$$

Note that we expect that the product of these three matrices

$$[6 \times 3][3 \times 3][3 \times 6] \text{ will yield a } [6 \times 6] \text{ matrix.}$$

We begin by multiplying the first two matrices

$$\begin{bmatrix} -1 & 0 & -2 \\ 0 & -2 & -1 \\ 1 & 0 & 0 \\ 0 & 0 & 1 \\ 0 & 0 & 2 \\ 0 & 2 & 0 \end{bmatrix} \begin{bmatrix} 1 & \frac{1}{3} & 0 \\ \frac{1}{3} & 1 & 0 \\ 0 & 0 & \frac{1}{3} \end{bmatrix} = \begin{bmatrix} -1 & -\frac{1}{3} & -\frac{2}{3} \\ -\frac{2}{3} & -2 & -\frac{1}{3} \\ 1 & \frac{1}{3} & 0 \\ 0 & 0 & \frac{1}{3} \\ 0 & 0 & \frac{2}{3} \\ \frac{2}{3} & 2 & 0 \end{bmatrix}$$

We recall the process of matrix multiplication defined in Section 2.2. For example the first term in the first row of the product matrix is

$$[(-1 \times 1) + (0 \times 1/3) + (-2 \times 0)] = -1$$

the second term in the first row of the product matrix is

$$[(-1 \times 1/3) + (0 \times 1) + (-2 \times 0)] = -1/3$$

and so on. Next we multiply

$$\begin{bmatrix} -1 & -\frac{1}{3} & -\frac{2}{3} \\ -\frac{2}{3} & -2 & -\frac{1}{3} \\ 1 & \frac{1}{3} & 0 \\ 0 & 0 & \frac{1}{3} \\ 0 & 0 & \frac{2}{3} \\ \frac{2}{3} & 2 & 0 \end{bmatrix} \begin{bmatrix} -1 & 0 & 1 & 0 & 0 & 0 \\ 0 & -2 & 0 & 0 & 0 & 2 \\ -2 & -1 & 0 & 1 & 2 & 0 \end{bmatrix} = \begin{bmatrix} \frac{7}{3} & \frac{4}{3} & -1 & -\frac{2}{3} & -\frac{4}{3} & -\frac{2}{3} \\ \frac{4}{3} & \frac{13}{3} & -\frac{2}{3} & -\frac{1}{3} & -\frac{2}{3} & -4 \\ -1 & -\frac{2}{3} & 1 & 0 & 0 & \frac{2}{3} \\ -\frac{2}{3} & -\frac{1}{3} & 0 & \frac{1}{3} & \frac{2}{3} & 0 \\ -\frac{4}{3} & -\frac{2}{3} & 0 & \frac{2}{3} & \frac{4}{3} & 0 \\ -\frac{2}{3} & -4 & \frac{2}{3} & 0 & 0 & 4 \end{bmatrix}$$

75

so $[B_1]^T[D][B_1]$ is

$$\frac{33.75 \times 10^6}{4}
\begin{bmatrix}
\frac{7}{3} & \frac{4}{3} & -1 & -\frac{2}{3} & -\frac{4}{3} & -\frac{2}{3} \\
\frac{4}{3} & \frac{13}{3} & -\frac{2}{3} & -\frac{1}{3} & -\frac{2}{3} & -4 \\
-1 & -\frac{2}{3} & 1 & 0 & 0 & \frac{2}{3} \\
-\frac{2}{3} & -\frac{1}{3} & 0 & \frac{1}{3} & \frac{2}{3} & 0 \\
-\frac{4}{3} & -\frac{2}{3} & 0 & \frac{2}{3} & \frac{4}{3} & 0 \\
-\frac{2}{3} & -4 & \frac{2}{3} & 0 & 0 & 4
\end{bmatrix}$$

and the stiffness matrix for element (1) is

$$[K_1] = 0.1 \times 1.0 \times \frac{33.75 \times 10^6}{4}
\begin{bmatrix}
\frac{7}{3} & \frac{4}{3} & -1 & -\frac{2}{3} & -\frac{4}{3} & -\frac{2}{3} \\
\frac{4}{3} & \frac{13}{3} & -\frac{2}{3} & -\frac{1}{3} & -\frac{2}{3} & -4 \\
-1 & -\frac{2}{3} & 1 & 0 & 0 & \frac{2}{3} \\
-\frac{2}{3} & -\frac{1}{3} & 0 & \frac{1}{3} & \frac{2}{3} & 0 \\
-\frac{4}{3} & -\frac{2}{3} & 0 & \frac{2}{3} & \frac{4}{3} & 0 \\
-\frac{2}{3} & -4 & \frac{2}{3} & 0 & 0 & 4
\end{bmatrix}$$

$$\text{Or } [K_1] = 8.437 \times 10^5
\begin{bmatrix}
\frac{7}{3} & \frac{4}{3} & -1 & -\frac{2}{3} & -\frac{4}{3} & -\frac{2}{3} \\
\frac{4}{3} & \frac{13}{3} & -\frac{2}{3} & -\frac{1}{3} & -\frac{2}{3} & -4 \\
-1 & -\frac{2}{3} & 1 & 0 & 0 & \frac{2}{3} \\
-\frac{2}{3} & -\frac{1}{3} & 0 & \frac{1}{3} & \frac{2}{3} & 0 \\
-\frac{4}{3} & -\frac{2}{3} & 0 & \frac{2}{3} & \frac{4}{3} & 0 \\
-\frac{2}{3} & -4 & \frac{2}{3} & 0 & 0 & 4
\end{bmatrix}$$

Note that this stiffness matrix is symmetrical.

The relation between the nodal displacements and the nodal forces is given by the matrix equation

$$[K_1]\{u\} = \{F\}$$

substituting the values for element (1)

$$
\begin{array}{cccccc}
u_1 & v_1 & u_2 & v_2 & u_3 & v_3
\end{array}
$$

$$
\begin{bmatrix}
\frac{7}{3} & \frac{4}{3} & -1 & -\frac{2}{3} & -\frac{4}{3} & -\frac{2}{3} \\
\frac{4}{3} & \frac{13}{3} & -\frac{2}{3} & -\frac{1}{3} & -\frac{2}{3} & -4 \\
-1 & -\frac{2}{3} & 1 & 0 & 0 & \frac{2}{3} \\
-\frac{2}{3} & -\frac{1}{3} & 0 & \frac{1}{3} & \frac{2}{3} & 0 \\
-\frac{4}{3} & -\frac{2}{3} & 0 & \frac{2}{3} & \frac{4}{3} & 0 \\
-\frac{2}{3} & -4 & \frac{2}{3} & 0 & 0 & 4
\end{bmatrix}
\begin{Bmatrix}
u_1 \\ v_1 \\ u_2 \\ v_2 \\ u_3 \\ v_3
\end{Bmatrix}
=
\begin{Bmatrix}
F_{u1} \\ F_{v1} \\ F_{u2} \\ F_{v2} \\ F_{u3} \\ F_{v3}
\end{Bmatrix}
=
\begin{Bmatrix}
0 \\ 0 \\ -300 \\ -400 \\ 0 \\ 0
\end{Bmatrix}
\frac{1}{8.437 \times 10^5}.
\tag{6.2.1}
$$

This completes the stiffness matrix equation for element (1).

6.2.2   STIFFNESS MATRIX FOR ELEMENT (2)

The element and its applied load are shown in Fig. 6.13

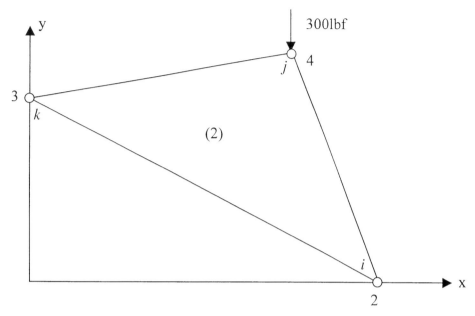

Fig. 6.13

The element data are collected in the following table.

Node	x	y	u	v	F_u	F_v
i (2)	2	0	u_2	v_2	F_{u2}	F_{v2}
j (4)	3/2	3/2	u_4	v_4	0	-300
k (3)	0	1	0	0	F_{u3}	F_{v3}

The element area is 1.25 in^2.
The matrix $[B_2]$ is:

$$\frac{1}{2 \times 1.25}\begin{bmatrix} \frac{1}{2} & 0 & 1 & 0 & -\frac{3}{2} & 0 \\ 0 & -\frac{3}{2} & 0 & 2 & 0 & -\frac{1}{2} \\ -\frac{3}{2} & \frac{1}{2} & 2 & 1 & -\frac{1}{2} & -\frac{3}{2} \end{bmatrix}$$

So $[B_2]^T [D] [B_2]$ is

$$\frac{33.75 \times 10^6}{4 \times 1.25^2}\begin{bmatrix} \frac{1}{2} & 0 & -\frac{3}{2} \\ 0 & -\frac{3}{2} & \frac{1}{2} \\ 1 & 0 & 2 \\ 0 & 2 & 1 \\ -\frac{3}{2} & 0 & -\frac{1}{2} \\ 0 & -\frac{1}{2} & -\frac{3}{2} \end{bmatrix}\begin{bmatrix} 1 & \frac{1}{3} & 0 \\ \frac{1}{3} & 1 & 0 \\ 0 & 0 & \frac{1}{3} \end{bmatrix}\begin{bmatrix} \frac{1}{2} & 0 & 1 & 0 & -\frac{3}{2} & 0 \\ 0 & -\frac{3}{2} & 0 & 2 & 0 & -\frac{1}{2} \\ -\frac{3}{2} & \frac{1}{2} & 2 & 1 & -\frac{1}{2} & -\frac{3}{2} \end{bmatrix}$$

First multiply

$$
\begin{bmatrix}
\frac{1}{2} & 0 & -\frac{3}{2} \\
0 & -\frac{3}{2} & \frac{1}{2} \\
1 & 0 & 2 \\
0 & 2 & 1 \\
-\frac{3}{2} & 0 & -\frac{1}{2} \\
0 & -\frac{1}{2} & -\frac{3}{2}
\end{bmatrix}
\begin{bmatrix}
1 & \frac{1}{3} & 0 \\
\frac{1}{3} & 1 & 0 \\
0 & 0 & \frac{1}{3}
\end{bmatrix}
=
\begin{bmatrix}
\frac{1}{2} & \frac{1}{6} & -\frac{1}{2} \\
-\frac{1}{2} & -\frac{3}{2} & \frac{1}{6} \\
1 & \frac{1}{3} & \frac{2}{3} \\
\frac{2}{3} & 2 & \frac{1}{3} \\
-\frac{3}{2} & -\frac{1}{2} & -\frac{1}{6} \\
-\frac{1}{6} & -\frac{1}{2} & -\frac{1}{2}
\end{bmatrix}
$$

Then multiply

$$
\begin{bmatrix}
\frac{1}{2} & \frac{1}{6} & -\frac{1}{2} \\
-\frac{1}{2} & -\frac{3}{2} & \frac{1}{6} \\
1 & \frac{1}{3} & \frac{2}{3} \\
\frac{2}{3} & 2 & \frac{1}{3} \\
-\frac{3}{2} & -\frac{1}{2} & -\frac{1}{6} \\
-\frac{1}{6} & -\frac{1}{2} & -\frac{1}{2}
\end{bmatrix}
\begin{bmatrix}
\frac{1}{2} & 0 & 1 & 0 & -\frac{3}{2} & 0 \\
0 & -\frac{3}{2} & 0 & 2 & 0 & -\frac{1}{2} \\
-\frac{3}{2} & \frac{1}{2} & 2 & 1 & -\frac{1}{2} & -\frac{3}{2}
\end{bmatrix}
=
\begin{bmatrix}
1 & -\frac{1}{2} & -\frac{1}{2} & -\frac{1}{6} & -\frac{1}{2} & \frac{2}{3} \\
-\frac{1}{2} & \frac{7}{3} & -\frac{1}{6} & -\frac{17}{6} & \frac{2}{3} & \frac{1}{2} \\
-\frac{1}{2} & -\frac{1}{6} & \frac{7}{3} & \frac{4}{3} & -\frac{11}{6} & -\frac{7}{6} \\
-\frac{1}{6} & -\frac{17}{6} & \frac{4}{3} & \frac{13}{3} & -\frac{7}{6} & -\frac{3}{2} \\
-\frac{1}{2} & \frac{2}{3} & -\frac{11}{6} & -\frac{7}{6} & \frac{7}{3} & \frac{1}{2} \\
\frac{2}{3} & \frac{1}{2} & -\frac{7}{6} & -\frac{3}{2} & \frac{1}{2} & 1
\end{bmatrix}
$$

So $[B_2]^T [D] [B_2]$ is

$$
\frac{33.75 \times 10^6}{4 \times 1.25^2}
\begin{bmatrix}
1 & -\frac{1}{2} & -\frac{1}{2} & -\frac{1}{6} & -\frac{1}{2} & \frac{2}{3} \\
-\frac{1}{2} & \frac{7}{3} & -\frac{1}{6} & -\frac{17}{6} & \frac{2}{3} & \frac{1}{2} \\
-\frac{1}{2} & -\frac{1}{6} & \frac{7}{3} & \frac{4}{3} & -\frac{11}{6} & -\frac{7}{6} \\
-\frac{1}{6} & -\frac{17}{6} & \frac{4}{3} & \frac{13}{3} & -\frac{7}{6} & -\frac{3}{2} \\
-\frac{1}{2} & \frac{2}{3} & -\frac{11}{6} & -\frac{7}{6} & \frac{7}{3} & \frac{1}{2} \\
\frac{2}{3} & \frac{1}{2} & -\frac{7}{6} & -\frac{3}{2} & \frac{1}{2} & 1
\end{bmatrix}
$$

and $[K_2] = t.\Delta. \, [B_2] [D] [B_2] =$

$$
0.1 \times 1.25 \times \frac{33.75 \times 10^6}{4 \times 1.25^2}
\begin{bmatrix}
1 & -\frac{1}{2} & -\frac{1}{2} & -\frac{1}{6} & -\frac{1}{2} & \frac{2}{3} \\
-\frac{1}{2} & \frac{7}{3} & -\frac{1}{6} & -\frac{17}{6} & \frac{2}{3} & \frac{1}{2} \\
-\frac{1}{2} & -\frac{1}{6} & \frac{7}{3} & \frac{4}{3} & -\frac{11}{6} & -\frac{7}{6} \\
-\frac{1}{6} & -\frac{17}{6} & \frac{4}{3} & \frac{13}{3} & -\frac{7}{6} & -\frac{3}{2} \\
-\frac{1}{2} & \frac{2}{3} & -\frac{11}{6} & -\frac{7}{6} & \frac{7}{3} & \frac{1}{2} \\
\frac{2}{3} & \frac{1}{2} & -\frac{7}{6} & -\frac{3}{2} & \frac{1}{2} & 1
\end{bmatrix}
$$

The stiffness matrix equation

$$
[K_2]\{u\} = \{F\}
$$

for element (2) is therefore

$$\frac{8.437 \times 10^5}{1.25}
\begin{bmatrix}
1 & -\frac{1}{2} & -\frac{1}{2} & -\frac{1}{6} & -\frac{1}{2} & \frac{2}{3} \\
-\frac{1}{2} & \frac{7}{3} & -\frac{1}{6} & -\frac{17}{6} & \frac{2}{3} & \frac{1}{2} \\
-\frac{1}{2} & -\frac{1}{6} & \frac{7}{3} & \frac{4}{3} & -\frac{11}{6} & -\frac{7}{6} \\
-\frac{1}{6} & -\frac{17}{6} & \frac{4}{3} & \frac{13}{3} & -\frac{7}{6} & -\frac{3}{2} \\
-\frac{1}{2} & \frac{2}{3} & -\frac{11}{6} & -\frac{7}{6} & \frac{7}{3} & \frac{1}{2} \\
\frac{2}{3} & \frac{1}{2} & -\frac{7}{6} & -\frac{3}{2} & \frac{1}{2} & 1
\end{bmatrix}
\begin{Bmatrix} u_2 \\ v_2 \\ u_4 \\ v_4 \\ u_3 \\ v_3 \end{Bmatrix}
= \begin{Bmatrix} F_{u2} \\ F_{v2} \\ F_{u4} \\ F_{v4} \\ F_{u3} \\ F_{v3} \end{Bmatrix}
= \begin{Bmatrix} 0 \\ 0 \\ 0 \\ -300 \\ 0 \\ 0 \end{Bmatrix}$$

(columns headed $u_2 \quad v_2 \quad u_4 \quad v_4 \quad u_3 \quad v_3$)

To assist the assembly process that follows, we make the multiplying factor in the equation the same as that for element (1) – that is 8.437×10^5. To this end we incorporate the factor $1/1.25$ within the stiffness matrix which we achieve by multiplying each term by $4/5$. The matrix equation for element (2) then becomes

$$\begin{bmatrix}
\frac{4}{5} & -\frac{2}{5} & -\frac{2}{5} & -\frac{2}{15} & -\frac{2}{5} & \frac{8}{15} \\
-\frac{2}{5} & \frac{28}{15} & -\frac{2}{15} & -\frac{34}{15} & \frac{8}{15} & \frac{2}{5} \\
-\frac{2}{5} & -\frac{2}{15} & \frac{28}{15} & \frac{16}{15} & -\frac{22}{15} & -\frac{14}{15} \\
-\frac{2}{15} & -\frac{34}{15} & \frac{16}{15} & \frac{52}{15} & -\frac{14}{15} & -\frac{6}{5} \\
-\frac{2}{5} & \frac{8}{15} & -\frac{22}{15} & -\frac{14}{15} & \frac{28}{15} & \frac{2}{5} \\
\frac{8}{15} & \frac{2}{5} & -\frac{14}{15} & -\frac{6}{5} & \frac{2}{5} & \frac{4}{5}
\end{bmatrix}
\begin{Bmatrix} u_2 \\ v_2 \\ u_4 \\ v_4 \\ u_3 \\ v_3 \end{Bmatrix}
= \begin{Bmatrix} 0 \\ 0 \\ 0 \\ -300 \\ 0 \\ 0 \end{Bmatrix} \frac{1}{8.437 \times 10^5} \qquad (6.2.2)$$

(columns headed $u_2 \quad v_2 \quad u_4 \quad v_4 \quad u_3 \quad v_3$)

We have headed each column in the element matrices with the nodal displacement component that each coefficient in that column multiplies. This will help in the next step in the analysis.

6.2.3 THE GLOBAL STIFFNESS MATRIX

We now form the Global Stiffness Matrix by ASSEMBLING the two element matrices. That matrix will be $[8 \times 8]$ because there are eight nodal displacements and correspondingly eight nodal force components, two for each of the four nodes. The matrix is formed by putting the coefficients from the stiffness matrices of the two elements in the appropriate row and column in the Global matrix. This process is called assembly. The frame for the Global matrix is displayed in Fig (6.14), which is an $[8 \times 8]$ table. At the top of each column is the nodal displacement that the coefficient in that column multiplies to get the equation for the force component. Each row is labelled with the nodal force component that the equation formed from that row yields. Note that the force and the displacement labels are not part of the stiffness matrix and the numbering of the rows and columns ignores them.

If we look at the stiffness matrix for element (1) eqn (6.2.1), the first row is the equation for F_{u1} so those coefficients are put in the first row of the Global matrix, which also corresponds to F_{u1}. The columns in the element (1) matrix correspond to u_1, v_1, u_2, v_2, u_3 and v_3. These are the first six columns in the Global matrix so we locate the element (1) coefficients there. Similarly, the second row of the element (1) matrix corresponds to the equation for F_{v1}, which occupies the second row of the Global matrix so the element (1) coefficients are put in this second row and again occupy

79

	u_1	v_1	u_2	v_2	u_3	v_3	u_4	v_4
F_{u1}	$\frac{7}{3}$	$\frac{4}{3}$	-1	$-\frac{2}{3}$	$-\frac{4}{3}$	$-\frac{2}{3}$	0	0
F_{v1}	$\frac{4}{3}$	$\frac{13}{3}$	$-\frac{2}{3}$	$-\frac{1}{3}$	$-\frac{2}{3}$	-4	0	0
F_{u2}	-1	$-\frac{2}{3}$	$1(+\frac{4}{5})$	$0(-\frac{2}{5})$	$0(-\frac{2}{5})$	$\frac{2}{3}(+\frac{8}{15})$	$(-\frac{2}{5})$	$(-\frac{2}{15})$
F_{v2}	$-\frac{2}{3}$	$-\frac{1}{3}$	$0(-\frac{2}{5})$	$\frac{1}{3}(+\frac{28}{15})$	$\frac{2}{3}(+\frac{8}{15})$	$0(+\frac{2}{5})$	$(-\frac{2}{15})$	$(-\frac{34}{15})$
F_{u3}	$-\frac{4}{3}$	$-\frac{2}{3}$	$0(-\frac{2}{5})$	$\frac{2}{3}(+\frac{8}{15})$	$\frac{4}{3}(+\frac{28}{15})$	$0(+\frac{2}{5})$	$(-\frac{22}{15})$	$(-\frac{14}{15})$
F_{v3}	$-\frac{2}{3}$	-4	$\frac{2}{3}(+\frac{8}{15})$	$0(+\frac{2}{5})$	$0(+\frac{2}{5})$	$4(+\frac{4}{5})$	$(-\frac{14}{15})$	$(-\frac{6}{5})$
F_{u4}	0	0	$(-\frac{2}{5})$	$(-\frac{2}{15})$	$(-\frac{22}{15})$	$(-\frac{14}{15})$	$(\frac{28}{15})$	$(\frac{16}{15})$
F_{v4}	0	0	$(-\frac{2}{15})$	$(-\frac{34}{15})$	$(-\frac{14}{15})$	$(-\frac{6}{5})$	$(\frac{6}{15})$	$(\frac{52}{15})$

Fig. 6.14

the first six columns. Correspondingly the 3rd 4th 5th and 6th rows of the element (1) matrix are transferred to the 3rd 4th 5th and 6th rows of the Global matrix.

We now add the coefficients of the element (2) matrix of eqn (6.2.2). Here the rows 1 to 6 are the equations for the nodal forces F_{u2}, F_{v2}, F_{u4}, F_{v4}, F_{u3} and F_{v3} which are rows 3, 4, 7, 8, 5 and 6 respectively of the Global matrix so the rows in the element matrix are not in the same order as those in the Global matrix and care is required to place the coefficients correctly. Further the coefficients in the six columns in the element (2) matrix multiply respectively the nodal displacements u_2, v_2, u_4, v_4, u_3 and v_3 which correspond to columns 3, 4, 7, 8, 5 and 6 of the Global matrix so the coefficients lie in a different order in the rows of the two matrices. The element (2) coefficients have been placed in brackets to distinguish them from the element (1) coefficients.

It might help to give an example of the transfer of a coefficient from the element (2) matrix to the Global matrix. The coefficient in the 3rd row and the 5th column of the element (2) matrix is $-22/15$. The row corresponds to F_{u4} and the column to u_3. The appropriate place for this coefficient in the Global matrix is row 7 (which is the equation for F_{u4}) and column 5 (which corresponds to u_3). It will be seen that the coefficient $-22/15$ indeed occupies that location in the Global matrix.

We add together the terms in each location in Fig. 6.14 to yield the following Global stiffness matrix

$$
\begin{array}{cccccccc}
u_1 & v_1 & u_2 & v_2 & u_3 & v_3 & u_4 & v_4
\end{array}
$$

$$
\begin{bmatrix}
\frac{7}{3} & \frac{4}{3} & -1 & -\frac{2}{3} & -\frac{4}{3} & -\frac{2}{3} & 0 & 0 \\
\frac{4}{3} & \frac{13}{3} & -\frac{2}{3} & -\frac{1}{3} & -\frac{2}{3} & -4 & 0 & 0 \\
-1 & -\frac{2}{3} & \frac{9}{5} & -\frac{2}{5} & -\frac{2}{5} & \frac{6}{5} & -\frac{2}{5} & -\frac{2}{15} \\
-\frac{2}{3} & -\frac{1}{3} & -\frac{2}{5} & \frac{11}{5} & \frac{6}{5} & \frac{2}{5} & -\frac{2}{15} & -\frac{34}{15} \\
-\frac{4}{3} & -\frac{2}{3} & -\frac{2}{5} & \frac{6}{5} & \frac{16}{5} & \frac{2}{5} & -\frac{22}{15} & -\frac{14}{15} \\
-\frac{2}{3} & -4 & \frac{6}{5} & \frac{2}{5} & \frac{2}{5} & \frac{4}{5} & -\frac{14}{15} & -\frac{6}{5} \\
0 & 0 & -\frac{2}{5} & -\frac{2}{15} & -\frac{22}{15} & -\frac{14}{15} & \frac{28}{15} & \frac{16}{15} \\
0 & 0 & -\frac{2}{15} & -\frac{34}{15} & -\frac{14}{15} & -\frac{6}{5} & \frac{16}{15} & \frac{52}{15}
\end{bmatrix}
\begin{Bmatrix}
u_1 \\ v_1 \\ u_2 \\ v_2 \\ u_3 \\ v_3 \\ u_4 \\ v_4
\end{Bmatrix}
=
\begin{Bmatrix}
F_{u1} \\ F_{v1} \\ -300 \\ -400 \\ F_{u3} \\ F_{v3} \\ 0 \\ -300
\end{Bmatrix}
\frac{1}{8.437 \times 10^5}
\qquad (6.2.3)
$$

Note that there are eight coefficients that are zero. These are always a feature of the global matrix because, as we have explained in Section 3.2, in a mesh of elements there will be nodes which, when displaced with all other nodes fixed, do not transmit a force to some of the other nodes in the system. Note too that there are no zeros in the element stiffness matrices because in the single element all nodes when displaced transmit forces to all the other nodes in the element.

6.2.4 THE CONDENSED STIFFNESS MATRIX AND ITS INVERSION

Our first objective is to derive values for the four unknown nodal displacements u_2, v_2, u_4 and v_4. For this we need four equations. Now we do not at this stage know the force components F_{u1}, F_{v1}, F_{u3} and F_{v3}, which are the forces needed to fix nodes 1 and 3. These equations are in rows 1, 2, 5 and 6 of the Global matrix. If we delete these rows, we are left with four equations for the forces we do know, that is F_{u2}, F_{v2}, F_{u4} and F_{v4}, lying in rows 3, 4, 7 and 8. So the only unknowns are the four nodal displacements u_2, v_2, u_4 and v_4.

Further, since we know that the displacements u_1 v_1 u_3 and v_3 are all zero, then the coefficients that multiply these displacements, which are in columns 1, 2, 5 and 6, will generate zero terms in the force equation. We therefore lose nothing by deleting the coefficients in these columns.

These musings lead to the formation of the condensed or reduced stiffness matrix, which is formed in this case by deleting rows 1, 2, 5 and 6 and columns 1, 2, 5 and 6 from the Global matrix. The result of this operation is

$$
\begin{array}{cccc}
u_2 & v_2 & u_4 & v_4
\end{array}
$$

$$
\begin{bmatrix}
\frac{9}{5} & -\frac{2}{5} & -\frac{2}{5} & -\frac{2}{15} \\
-\frac{2}{5} & \frac{11}{5} & -\frac{2}{5} & -\frac{34}{15} \\
-\frac{2}{5} & -\frac{2}{15} & \frac{28}{15} & \frac{16}{15} \\
-\frac{2}{15} & -\frac{34}{15} & \frac{16}{15} & \frac{52}{15}
\end{bmatrix}
\begin{Bmatrix}
u_2 \\ v_2 \\ u_4 \\ v
\end{Bmatrix}
=
\begin{Bmatrix}
-300 \\ -400 \\ 0 \\ -300
\end{Bmatrix}
\frac{1}{8.437 \times 10^5}
$$

We tidy this matrix equation up a bit to make it more streamlined for further analysis. We take out the common factor 1/15 from the matrix and incorporate it with the other constant into the force vector on the right-hand side. The final form of the matrix equation is then

$$
\begin{bmatrix}
27 & -6 & -6 & -2 \\
-6 & 33 & -2 & -34 \\
-6 & -2 & 28 & 16 \\
-2 & -34 & 16 & 52
\end{bmatrix}
\begin{Bmatrix} u_2 \\ v_2 \\ u_4 \\ v_4 \end{Bmatrix}
=
\begin{Bmatrix} -5.338 \times 10^{-3} \\ -7.117 \times 10^{-3} \\ 0 \\ -5.338 \times 10^{-3} \end{Bmatrix}
$$

This represents four equations with only four unknowns, which are the required nodal displacements, so the problem is solved. We extract the values of the four displacements by inverting the matrix, which we shall do using Gaussian elimination, which we explained in Section 2.4.2. The objective is to modify the matrix to generate zeros in all the terms below the leading diagonal. We now proceed to achieve that. For clarity we begin by copying the matrix equation which is our starting point

$$
\begin{bmatrix}
27 & -6 & -6 & -2 \\
-6 & 33 & -2 & -34 \\
-6 & -2 & 28 & 16 \\
-2 & -34 & 16 & 52
\end{bmatrix}
\begin{Bmatrix} u_2 \\ v_2 \\ u_4 \\ v_4 \end{Bmatrix}
=
\begin{Bmatrix} -5.338 \times 10^{-3} \\ -7.117 \times 10^{-3} \\ 0 \\ -5.338 \times 10^{-3} \end{Bmatrix}
$$

We multiply row 2 by 27/6 and add row 1 to it. Note that row 2 includes both the second row of the $[4 \times 4]$ matrix and also the second row of the force vector on the right-hand side (-7.117×10^{-3}). By multiplying both of these by the same factor the balance of the equation that these terms represent is unchanged. Also by adding row 1, which again is a balanced equation, to the result we are again adding an equal quantity to both sides of the equation and again the balance is preserved.

Let us explain the convention we will adopt to display the progressive modification of the matrix equation by taking the first step as an example.

The result of multiplying row 2 by 27/6 is given in brackets in the following matrix equation. The result of adding row 1 to this is displayed on the row below it. This becomes the new row 2 for the next step.

$$
\begin{bmatrix}
27 & -6 & -6 & -2 \\
(-27) & (148.5) & (-9) & (-153) \\
0 & 142.5 & -15 & -155 \\
-6 & -2 & 28 & 16 \\
-2 & -34 & 16 & 52
\end{bmatrix}
\begin{Bmatrix} u_2 \\ \\ v_2 \\ u_4 \\ v_4 \end{Bmatrix}
=
\begin{Bmatrix} -5.338 \times 10^{-3} \\ (-32.027 \times 10^{-3}) \\ -37.365 \times 10^{-3} \\ 0 \\ -5.228 \times 10^{-3} \end{Bmatrix}
$$

Now multiply row 3 by 27/6 and add row 1 to it. To emphasise the convention, note that the 3rd row, in brackets, is the result of multiplying row 3 by 27/6 and the row below it is the result of adding row 1 to it. This now becomes the new row 3.

$$
\begin{bmatrix}
27 & -6 & -6 & -2 \\
0 & 142.5 & -15 & -155 \\
(-27) & (-9) & (126) & (72) \\
0 & -15 & 120 & 70 \\
-2 & -34 & 16 & 52
\end{bmatrix}
\begin{Bmatrix} u_2 \\ \\ v_2 \\ u_4 \\ v_4 \end{Bmatrix}
=
\begin{Bmatrix} -5.338 \times 10^{-3} \\ -37.365 \times 10^{-3} \\ (0) \\ -5.338 \times 10^{-3} \\ -5.338 \times 10^{-3} \end{Bmatrix}
$$

82

Now multiply row 4 by 27/2 and add row 1 to it

$$
\begin{bmatrix}
27 & -6 & -6 & -2 \\
0 & 142.5 & -15 & -155 \\
0 & -15 & 120 & 70 \\
(-27) & (-459) & (216) & (702) \\
0 & -465 & 210 & 700
\end{bmatrix}
\begin{Bmatrix}
u_2 \\
\\
v_2 \\
u_4 \\
v_4
\end{Bmatrix}
=
\begin{Bmatrix}
-5.338 \times 10^{-3} \\
-37.365 \times 10^{-3} \\
-5.338 \times 10^{-3} \\
(-72.063 \times 10^{-3)} \\
-77.401 \times 10^{-3}
\end{Bmatrix}
$$

Multiply row 3 by 142.5/15 and add row 2 to it

$$
\begin{bmatrix}
27 & -6 & -6 & -2 \\
0 & 142.5 & -15 & -155 \\
(0) & (-142.5) & (1140) & (665) \\
0 & 0 & 1125 & 510 \\
0 & -465 & 210 & 700
\end{bmatrix}
\begin{Bmatrix}
u_2 \\
\\
v_2 \\
u_4 \\
v_4
\end{Bmatrix}
=
\begin{Bmatrix}
-5.338 \times 10^{-3} \\
-37.365 \times 10^{-3} \\
(-50.711 \times 10^{-3}) \\
-88.076 \times 10^{-3} \\
-77.401 \times 10^{-3}
\end{Bmatrix}
$$

Multiply row 4 by 142.4/465 and add row 2 to it

$$
\begin{bmatrix}
27 & -6 & -6 & -2 \\
0 & 142.5 & -15 & -155 \\
0 & 0 & 1125 & 510 \\
(0) & (-142.5) & (64.3548) & (214.516) \\
0 & 0 & 49.355 & 59.516
\end{bmatrix}
\begin{Bmatrix}
u_2 \\
\\
v_2 \\
u_4 \\
v_4
\end{Bmatrix}
=
\begin{Bmatrix}
-5.338 \times 10^{-3} \\
-37.365 \times 10^{-3} \\
-88.076 \times 10^{-3} \\
(-23.720 \times 10^{-3}) \\
-61.085 \times 10^{-3}
\end{Bmatrix}
$$

Multiply row 4 by 1125/49.355 and subtract row 3 from it

$$
\begin{bmatrix}
27 & -6 & -6 & -2 \\
0 & 142.5 & -15 & -155 \\
0 & 0 & 1125 & 510 \\
(0) & (0) & (1125) & (1356.61) \\
0 & 0 & 0 & 846.61
\end{bmatrix}
\begin{Bmatrix}
u_2 \\
\\
v_2 \\
u_4 \\
v_4
\end{Bmatrix}
=
\begin{Bmatrix}
-5.338 \times 10^{-3} \\
-37.365 \times 10^{-3} \\
-88.076 \times 10^{-3} \\
(-1392.374 \times 10^{-3}) \\
-1304.298 \times 10^{-3}
\end{Bmatrix}
$$

We have now achieved our objective of modifying the matrix to produce zero terms below the main diagonal. The final form of the matrix equation is

$$
\begin{bmatrix}
27 & -6 & -6 & -2 \\
0 & 142.5 & -15 & -155 \\
0 & 0 & 1125 & 510 \\
0 & 0 & 0 & 846.61
\end{bmatrix}
\begin{Bmatrix}
u_2 \\
v_2 \\
u_4 \\
v_4
\end{Bmatrix}
=
\begin{Bmatrix}
-5.338 \times 10^{-3} \\
-37.365 \times 10^{-3} \\
-88.076 \times 10^{-3} \\
-1304.298 \times 10^{-3}
\end{Bmatrix}
$$

We multiply out the 4th row, ignoring the zero terms

$$846.61 v_4 = -1304.298 \times 10^{-3}$$

$$v_4 = -0.0015406 \text{ in}$$

Multiply out the 3rd row

$$1124 u_4 + 510 v_4 = -88.076 \times 10^{-3}$$

so

$$u_4 = \frac{1}{1125}[-88.076 \times 10^{-3} - 510 \times (-0.0015406)]$$

and

$$u_4 = +0.0006201 \text{ in}$$

Multiply out the 2nd row

$$142.5 v_2 - 15 u_4 - 155 v_4 = -37.365 \times 10^{-3}$$

or

$$v_2 = \frac{1}{142.5}[-37.365 \times 10^{-3} + 15 \times 0.0006201 + 155 \times (-0.0015406)]$$

and

$$v_2 = -0.0018727 \text{ in}$$

Multiply out the 1st row

$$27 u_2 - 6 v_2 - 6 u_4 - 2 v_4 = -5.338 \times 10^{-3}$$

so

$$u_2 = \frac{1}{27}[-5.338 \times 10^{-3} + 6 \times (-0.0018727) + 6 \times 0.0006201 + 2 \times (-0.0015406)]$$

and

$$u_2 = -0.00059018 \text{ in}$$

6.2.5 CALCULATION OF THE ELEMENT STRESSES

The stresses are given by the basic relation

$$\{\sigma\} = [D]\{\varepsilon\}$$

where σ is the stress vector, [D] is the elastic constants matrix and ε is the strain vector. We know that

$$\{\varepsilon\} = [B]\{u\}$$

where u is the nodal displacement vector, so

$$\{\sigma\} = [D][B]\{u\} \tag{6.2.4}$$

We have established all the matrices on the right-hand side of this equation so we can easily calculate the stresses.

Stresses in element (1)
Substituting in equation (6.2.4)

$$\begin{Bmatrix} \sigma_{xx} \\ \sigma_{yy} \\ \sigma_{xy} \end{Bmatrix} = \frac{33.75 \times 10^6}{2} \begin{bmatrix} 1 & \frac{1}{3} & 0 \\ \frac{1}{3} & 1 & 0 \\ 0 & 0 & \frac{1}{3} \end{bmatrix} \begin{bmatrix} -1 & 0 & 1 & 0 & 0 & 0 \\ 0 & -2 & 0 & 0 & 0 & 2 \\ -2 & -1 & 0 & 1 & 2 & 0 \end{bmatrix} \begin{Bmatrix} u_1 \\ v_1 \\ u_2 \\ v_2 \\ u_3 \\ v_3 \end{Bmatrix}$$

so

$$\begin{Bmatrix} \sigma_{xx} \\ \sigma_{yy} \\ \sigma_{xy} \end{Bmatrix} = 16.87 \times 10^6 \begin{bmatrix} -1 & -\frac{2}{3} & 1 & 0 & 0 & \frac{2}{3} \\ -\frac{1}{3} & -2 & \frac{1}{3} & 0 & 0 & 2 \\ -\frac{2}{3} & -\frac{1}{3} & 0 & \frac{1}{3} & \frac{2}{3} & 0 \end{bmatrix} \begin{Bmatrix} 0 \\ 0 \\ -0.00059018 \\ -0.0018727 \\ 0 \\ 0 \end{Bmatrix}$$

Multiplying out the three rows of the equations, ignoring zero terms, gives:

Row 1
$$\sigma_{xx} = 16.87 \times 10^6 \times 1 \times (-0.00059018)$$
$$= -9955 \, \text{lbf in}^{-2}$$

Row 2
$$\sigma_{yy} = 16.87 \times 10^6 \times 1/3 \times (-0.00059018)$$
$$= -3318 \, \text{lbf in}^{-2}$$

Row 3
$$\sigma_{xy} = 16.87 \times 10^6 \times 1/3 \times (-0.0018727)$$
$$= -10531 \, \text{lbf in}^{-2}$$

Stresses in Element (2)

$$\begin{Bmatrix} \sigma_{xx} \\ \sigma_{yy} \\ \sigma_{xy} \end{Bmatrix} = \frac{16.87 \times 10^6}{1.25} \begin{bmatrix} 1 & \frac{1}{3} & 0 \\ \frac{1}{3} & 1 & 0 \\ 0 & 0 & \frac{1}{3} \end{bmatrix} \begin{bmatrix} \frac{1}{2} & 0 & 1 & 0 & -\frac{3}{2} & 0 \\ 0 & -\frac{3}{2} & 0 & 2 & 0 & -\frac{1}{2} \\ -\frac{3}{2} & \frac{1}{2} & 2 & 1 & -\frac{1}{2} & -\frac{3}{2} \end{bmatrix} \begin{Bmatrix} u_2 \\ v_2 \\ u_4 \\ v_4 \\ u_3 \\ v_3 \end{Bmatrix}$$

or

$$\begin{Bmatrix} \sigma_{xx} \\ \sigma_{yy} \\ \sigma_{xy} \end{Bmatrix} = \frac{16.87 \times 10^6}{1.25} \begin{bmatrix} \frac{1}{2} & -\frac{1}{2} & 1 & \frac{2}{3} & -\frac{3}{2} & -\frac{1}{6} \\ \frac{1}{6} & -\frac{3}{2} & \frac{1}{3} & 2 & -\frac{1}{2} & -\frac{1}{2} \\ -\frac{1}{2} & \frac{1}{6} & \frac{2}{3} & \frac{1}{3} & -\frac{1}{6} & -\frac{1}{2} \end{bmatrix} \begin{Bmatrix} -.00059018 \\ -.0018727 \\ 0.00062011 \\ -.0015406 \\ 0 \\ 0 \end{Bmatrix}$$

85

Multiplying out the rows to extract the stresses

Row 1

$$\sigma_{xx} = \frac{16.87 \times 10^6}{1.25}\left[-\frac{0.00059018}{2}+\frac{0.0018727}{2}+0.00062011-\frac{2}{3}\times 0.0015406\right]$$

$$= 3163\, \text{lbf in}^{-2}$$

Row 2

$$\sigma_{yy} = \frac{16.87x10^6}{1.25}\left[-\frac{0.00059018}{6}+\frac{3\times 0.0018727}{2}+\frac{0.00062011}{3}-2\times 0.0015406\right]$$

$$= -2211\, \text{lbf in}^{-2}$$

Row 3

$$\sigma_{xy} = \frac{16.87 \times 10^6}{1.25}\left[\frac{0.00059018}{2}-\frac{0.0018727}{6}+\frac{2\times 0.00062011}{3}-\frac{0.0015406}{3}\right]$$

$$= -1582\, \text{lbf in}^{-2}$$

Note that negative direct stresses are compressive.

A Test of Force Equilibrium
A useful test of the validity of a solution is to check that the calculated forces applied to the element are in static equilibrium– as they must be. There are many equilibrium conditions. The one we shall choose is that the resultant horizontal force (in the direction of x) on the element must be zero. That is

$$F_{u1} + F_{u2} + F_{u3} + F_{u4} = 0$$

We know that $F_{u2} = -300\,\text{lbf}$ and that $F_{u4} = 0$. Now since we now know all the nodal displacements, the uncondensed Global matrix (6.2.3) gives F_{u1} and F_{u3} explicitly. We now demonstrate this.

If we return to that matrix, eqn (6.2.3), we see that the equation for F_{u1} is the top row of the matrix equation. Multiplying this out, ignoring the zeros, we get

$$-1 \times u_2 - 2/3 \times v_2 = F_{u1} \times \frac{1}{8.437 \times 10^5}$$

Substituting the known values for the two nodal displacements

$$[-0.00059018 - 2/3 \times 0.001827] \times 8.437 \times 10^5 = F_{u1}$$

so

$$F_{u1} = 1549.98\,\text{lbf}$$

The equation for F_{u3} is row 5. Multiplying this out ignoring zeros yields

$$[-2/5 \times u_2 + 6/5 \times v_2 - 22/15 \times u_4 - 14/15 \times v_4] = F_{u3} \times \frac{1}{8.437x10^5}$$

substituting values for the four nodal displacements

$$[-2/5(-0.00059018) + 6/5(-0.0018727) - 22/15(0.0006201)$$
$$- 14/15(-0.0015406)] \times 8.437 \times 10^5 = F_{u3}$$

so

$$F_{u3} = -1249.96 \, \text{lbf}$$

and

$$F_{u1} + F_{u2} + F_{u3} + F_{u4} = 1549.98 - 300 + 0 - 1249.96$$
$$= 0.02 \, \text{lbf}$$

This is close enough to zero to encourage us to have faith in our solution.

6.2.6   Results from a Finite Element Package

It is of interest to compare the solution we have achieved with that obtained using one of the commercial finite element packages. To this end the problem was run on the LUSAS software again using triangular constant strain elements. The results were:

Displacements		Stresses	
$u_2 = -0.0005896$ in			
$v_2 = -0.001871$ in		Element (1)	Element (2)
$u_4 = +0.0006196$ in		$\sigma_{xx} = -9950 \, \text{lbf in}^{-2}$	$\sigma_{xx} = 3160 \, \text{lbf in}^{-2}$
$v_4 = -0.001549$ in		$\sigma_{yy} = -3316 \, \text{lbf in}^{-2}$	$\sigma_{yy} = -2210 \, \text{lbf in}^{-2}$
		$\sigma_{xy} = -10525 \, \text{lbf in}^{-2}$	$\sigma_{xy} = -1580 \, \text{lbf in}^{-2}$

These data are seen to be in excellent accord with the results we secured 'the hard way'.

We must not give much engineering credence to this very crude model of the real problem. There are however features which are consistent with reality. For example, it is evident that the loading system puts the plate predominantly in clockwise bending. So we would expect that the direct stress σ_{xx} would be tensile in the upper part of the plate and compressive in the lower part. This is the case – σ_{xx} is tensile in element (2) – the upper element and compressive in element (1) – the lower one.

6.2.7   A More Meaningful Solution using the LUSAS Package

A solution to the problem that we have analysed in this section using only two elements was obtained with the LUSAS package with a mesh comprising 380 elements. The mesh is displayed in Fig. 6.15 for exactly the same sheet geometry and applied loads as in Fig. 6.10 and is formed entirely of triangular constant strain elements. All the nodes along the left-hand edge of the plate are fixed.

Figure 6.16 is a graph of the calculated variation of the direct stress in the horizontal direction (σ_{xx}) along the constrained left-hand edge of the sheet. The origin for the x-axis is the bottom left-hand corner of the plate. The stress σ_{xx} varies almost linearly from compression at the bottom left-hand corner to tension at the top left-hand corner. This confirms our intuition that, along this edge, the effect of the loading system is to cause a predominantly bending stress to develop over this plane.

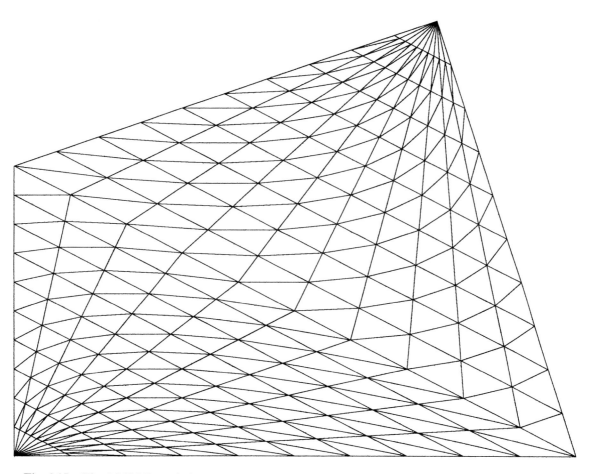

Fig 6.15 The LUSAS mesh for the loaded plate of Fig. 6.10. All the nodes along the left-hand vertical edge of the plate are fixed.

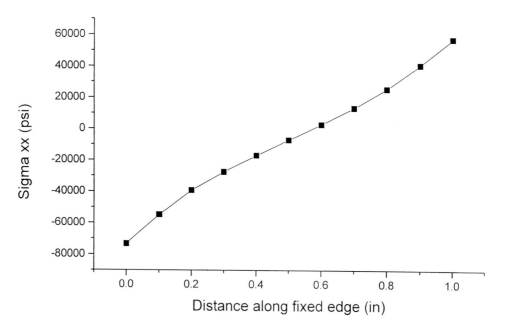

Fig. 6.16 Variation of the horizontal stress σ_{xx} along the fixed left-hand edge of the plate. The origin of the x-axis coincides with the bottom left-hand corner of the plate.

CHAPTER 7   FINITE ELEMENT ANALYSIS USING HIGHER ORDER ELEMENTS

The analysis which we have just completed using triangular elements assumes that the displacement within the element varies linearly across it. This yields strains and stresses that are constant throughout the element. To analyse steeply varying stress patterns using this element we need to divide the loaded body into a large number of elements in order to achieve the desirable mesh in which the stress difference between adjacent elements is small compared to the overall stress differences across the model. An equally accurate solution can be achieved using fewer elements, and less computing time, by adopting elements in which the displacement is assumed to vary with position in the element as a higher-than-linear order, for example as the square or as the cube of the position co-ordinates. Since the strains are the spatial derivatives of the displacement (see eqns 6.5, 6.6 and 6.7), then if the displacement varies as the square of the position, the strain will vary linearly so a linear variation of stress is built into the element. This allows us to secure a closer approximation to the actual stress pattern using less computer time than with elements over which the stress is constant.

We develop the rather complex analysis involved in the use of this type of element by choosing a quadrilateral element with eight nodes:

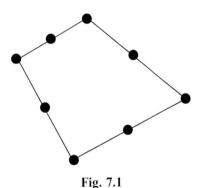

Fig. 7.1

Note that the element has nodes both at the corners and along the sides. The side nodes need not be at the mid-point.

A very important concept used in the analysis attached to this type of element is that of MAPPING. This involves starting with a simple form of element, in this case a square, and changing the shape of this by a geometrical transformation. The range of shapes that it is possible to generate is pictured in Fig. 7.2. So it is possible to generate any shape of quadrilateral with either straight or curved sides.

The square element is referred to as a SERENDIPITY ELEMENT or MASTER ELEMENT. Serendip is the ancient name for Ceylon which folklore asserts was discovered by chance. So a serendipity feature is one that arises by chance, and it is claimed that the discovery of the powerful potential of the serendipity element was of this essence.

The power of this concept lies in the ability to formulate all the finite element analysis for the simple serendipity element which then forms the generic analysis for the rich range of element shapes that can be derived from it. Then, to calculate the behaviour of the particular derived element, it turns out that all that is necessary is to substitute the nodal co-ordinates of the particular element into the generic analysis.

An important feature of the mapping process is the SHAPE FUNCTION. We need to introduce this before we can deal with mapping.

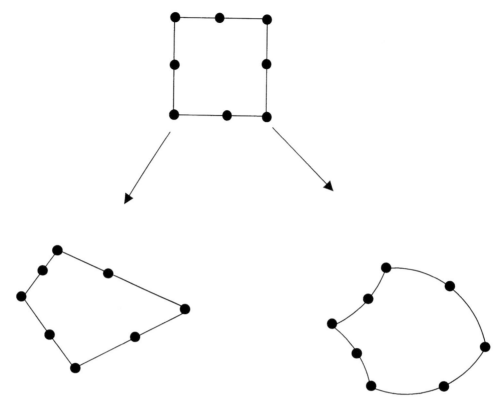

Fig. 7.2

7.1 SHAPE FUNCTIONS

We introduce SHAPE FUNCTIONS by looking at their simplest form in a one-dimensional line element along which the displacement varies linearly.

Figure 7.3 shows two ways of defining the co-ordinates of a point in the element, with nodes 1 and 2 at the ends. These co-ordinates would then be used to define the variation of some quantity along the element, for example the elastic displacement. We shall call this dependent variable ϕ.

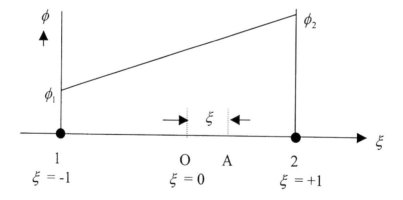

Fig. 7.3a   Serendipity or Master Element

In Fig. 7.3a we use a co-ordinate ξ (the Greek letter Xi) which is called an **INTRINSIC** or **NATURAL CO-ORDINATE**. It varies linearly from -1 at node 1 to $+1$ at node 2. We assume that the elastic particle displacement in the element ϕ varies linearly from ϕ_1 at node 1 to ϕ_2 at

90

node 2. This linear variation is called an INTERPOLATION FUNCTION. We call Fig. 7.3a a serendipity or master element.

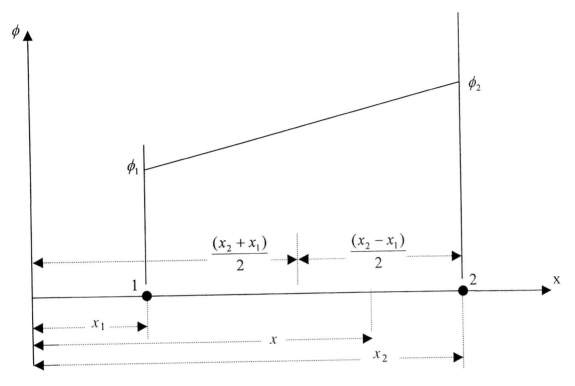

Fig. 7.3b The Real Element

In Fig. 7.3b, a point in the real element is defined by the co-ordinate x measured from a co-ordinate origin outside the element. Node 1 of the element lies at $x = x_1$ and node 2 at $x = x_2$. The mid-point of the element lies at $x = \frac{(x_2 + x_1)}{2}$ and half the length of the element is $\frac{(x_2 - x_1)}{2}$. Noting in Fig. 7.3a that $\xi = \frac{OA}{O2}$, it follows from Fig. 7.3b that

$$\xi = \frac{x - \frac{(x_2 + x_1)}{2}}{\frac{(x_2 - x_1)}{2}}. \tag{7.2}$$

We have assumed that the displacement ϕ varies linearly with ξ so we write

$$\phi = a_1 + a_2 \xi \tag{7.3}$$

At node 1 (Fig. 7.3a) $\phi = \phi_1$ and $\xi = -1$ so substituting in (7.3)

$$\phi_1 = a_1 - a_2 \tag{7.4}$$

At node 2

$$\phi = \phi_2 \quad \text{and} \quad \xi = +1$$

so

$$\phi_2 = a_1 + a_2 \tag{7.5}$$

91

Adding eqns (7.4) and (7.5)

$$\phi_1 + \phi_2 = 2a_1 \text{ or } a_1 = \frac{\phi_1 + \phi_2}{2}$$

Subtracting eqns (7.4) and (7.5)

$$\phi_2 - \phi_1 = 2a_2 \text{ or } a_2 = \frac{\phi_2 - \phi_1}{2}$$

Substituting the values of a_1 and a_2 in eqn (7.3)

$$\phi = \left(\frac{\phi_1 + \phi_2}{2}\right) + \left(\frac{\phi_2 - \phi_1}{2}\right)\xi$$

or rearranging the terms

$$\phi = \left(\frac{1-\xi}{2}\right)\phi_1 + \left(\frac{1+\xi}{2}\right)\phi_2 \tag{7.6}$$

$$= N_1\phi_1 \quad + \quad N_2\phi_2 \tag{7.7}$$

where N_1 and N_2 are called SHAPE FUNCTIONS.

$$N_1 = \left(\frac{1-\xi}{2}\right) \text{ and } N_2 = \left(\frac{1+\xi}{2}\right)$$

We plot these two functions against ξ in Fig. 7.4

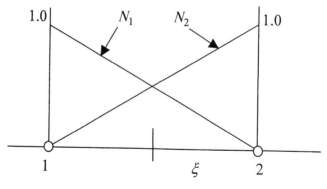

Fig. 7.4

The shape functions have the following characteristics

N_1 has a value of 1 at node 1 and falls linearly to zero at node 2.
N_2 has a value of 1 at node 2 and falls linearly to zero at node 1.

At any value of ξ,

$$N_1 + N_2 = \left(\frac{1-\xi}{2}\right) + \left(\frac{1+\xi}{2}\right) = 1$$

So the sum of the shape functions is unity.

The co-ordinates of a point in the element in Fig. 7.4a can be related to the co-ordinates of a point in Fig. 7.4b in the following way. Restating eqn (7.2)

$$\xi = \frac{x - \dfrac{x_2 + x_1}{2}}{\dfrac{x_2 - x_1}{2}}$$

rearranging (7.2)

$$\xi(x_2 - x_1) = 2x - (x_2 + x_1)$$

or

$$x = \left(\frac{1 - \xi}{2}\right)x_1 + \left(\frac{1 + \xi}{2}\right)x_2$$

or

$$x = N_1 x_1 + N_2 x_2 \tag{7.8}$$

N_1 and N_2 are the same shape functions that described the variation of displacement ϕ in the master element of Fig. 7.3a.

Equation (7.8) can be used to give the x co-ordinate of a point in the real element of Fig. 7.3b corresponding to a point in the master element of Fig. 7.3a defined by its ξ co-ordinate. So eqn (7.8) relates a point in (A) to a point in (B) and in this sense is a MAPPING RELATION in the same sense that a point on a map relates to a point in the real territory that the map represents.

An element like Fig. 7.3b for which the variation of displacement in the master element and the relation between the co-ordinates in the master and the real element are described by the same shape functions is called an ISOPARAMETRIC ELEMENT.

The mapping process becomes more interesting when we use higher order interpolations. Let us see how it works out using a quadratic rather than a linear variation. To do this we need to adopt a 3-node linear element. The master element and the variation of the displacement ϕ along it are shown in Fig. 7.5.

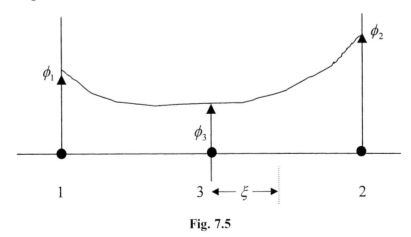

Fig. 7.5

The element has nodes 1 and 2 at the ends and node 3 in the middle. We assume that the displacement varies parabolically with ξ so

$$\phi = a_1 + a_2 \xi + a_3 \xi^2 \tag{7.9}$$

At $\qquad\qquad\qquad \xi = -1 \qquad \phi = \phi_1 \qquad$ so $\qquad \phi_1 = a_1 - a_2 + a_3$ $\qquad\qquad$ (7.10)

At $\qquad\qquad\qquad \xi = 0 \qquad \phi = \phi_3 \qquad$ so $\qquad \phi_3 = a_1$ $\qquad\qquad\qquad\quad$ (7.11)

At $\qquad\qquad\qquad \xi = +1 \qquad \phi = \phi_2 \qquad$ so $\qquad \phi_2 = a_1 + a_2 + a_3$ $\qquad\qquad$ (7.12)

Adding (7.10) and (7.12) to get a_3 and subtracting to get a_2

$$a_3 = \frac{\phi_1 + \phi_2 - 2\phi_3}{2} \qquad \text{and} \qquad a_2 = \frac{\phi_2 - \phi_1}{2} \qquad \text{and from (7.11) } a_1 = \phi_3$$

Substituting these values in (7.9) and rearranging

$$\phi = \{-\xi/2(1 - \xi)\}\phi_1 + \{\xi/2(1 + \xi)\}\phi_2 + (1 + \xi)(1 - \xi)\phi_3 \qquad (7.13)$$

or

$$\phi = N_1\phi_1 + N_2\phi_2 + N_3\phi_3 \qquad (7.14)$$

where N_1, N_2 and N_3 are the shape functions for this element. Their variation with ξ is shown in Fig. 7.6

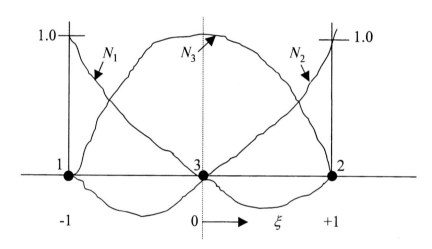

Fig. 7.6

At node 1 $N_1 = 1$ and $N_2 = N_3 = 0$

At node 2 $N_2 = 1$ and $N_1 = N_3 = 0$

At node 3 $N_3 = 1$ and $N_1 = N_2 = 0$

At any value of ξ, $N_1 + N_2 + N_3 = 1$

So the three shape functions for this 3-node master or serendipity element are

$$N_1 = -\xi/2(1 - \xi)$$
$$N_2 = \xi/2(1 + \xi)$$
$$N_3 = (1 + \xi)(1 - \xi) \qquad (7.15)$$

We will now show how to use these shape functions to derive a real isoparametric element from this master element. The real element is

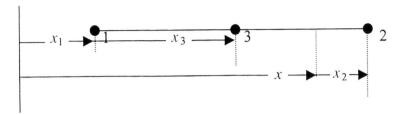

Fig. 7.7

We assert that the ξ co-ordinate in the master element is related to the x co-ordinate in the real element by

$$x = N_1 x_1 + N_2 x_2 + N_3 x_3 \qquad (7.16)$$

where x_1, x_2 and x_3 are the nodal co-ordinates and x is the co-ordinate of a point in the real element corresponding to the point ξ in the master element. The shape functions in (7.16) are exactly the same functions of ξ as those in eqn (7.15).

We will now work through s simple example to show how the mapping process works out in practice.

7.1.1 AN EXAMPLE OF MAPPING IN ONE DIMENSION

Let us map ξ points on a 3-node master element:

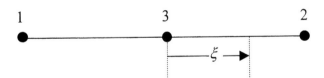

on to a real element whose nodal x co-ordinates are:

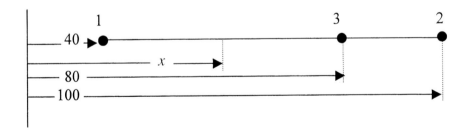

Fig. 7.8

Note that node 3 is not at the mid-point of nodes 1 and 2. It need not be.

The following table gives selected values of ξ and the calculated values of x using

$$x = N_1 x_1 + N_2 x_2 + N_3 x_3$$

and the values of N_1, N_2 and N_3 in eqn (7.15). The N values are also tabulated.

95

TABLE (6.1) Table Showing the Relation Between the Master Element Co-ordinates ξ and the Real Element Co-ordinates x for the Element in Fig. 7.8.

$$x_1 = 40 \quad x_2 = 100 \quad x_3 = 80$$

$$N_1 = -\xi/2(1-\xi) \quad N_2 = \xi/2(1+\xi) \quad N_3 = (1+\xi)(1-\xi)$$

ξ	N_1	N_2	N_3	x
1	0	1	0	100
0.75	−3/32	21/32	7/16	96.875
0.5	−1/8	3/8	3/4	92.5
0.25	−3/32	5/32	15/16	86.875
0	0	0	1	80
−0.25	5/32	−3/32	15/16	71.875
−0.5	3/8	−1/8	3/4	62.5
−0.75	21/32	−3/32	7/16	51.875
−1	1	0	0	40

In Fig. 7.9 the values of x are plotted against ξ. The relation between the two co-ordinates is evidently non-linear demonstrating that the mapping process is capable of producing a controlled distortion of the master element.

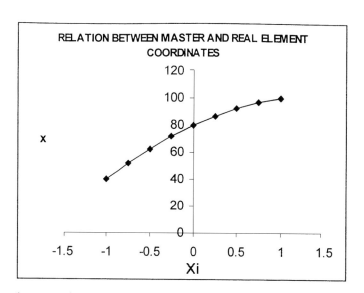

Fig. 7.9 Relation between the master element coordinate (Xi) and the real element coordinate (x) for the two linear elements in Fig. 7.8.

The capacity of mapping to produce a wide range of element shapes is even more convincingly demonstrated with two-dimensional elements. We will now present an example of this, which will provide helpful background to the latter section where we develop the use of quadrilateral elements for stress analysis.

7.1.2 MAPPING IN TWO DIMENSIONS

We take as our example the widely used 8-node element for which the master element is a square. The 8 nodes are numbered. We need a two-dimensional co-ordinate system for which we choose the Greek symbols (ξ, η) – which are Xi and Eta in line with their widespread usage in the finite element literature. The origin of this co-ordinate system is at the centre of the square element. As with the 3-node element ξ ranges from -1 to $+1$ as does η. The co-ordinates of nodes 1, 2, 3 and 4 are given in Fig. 7.10.

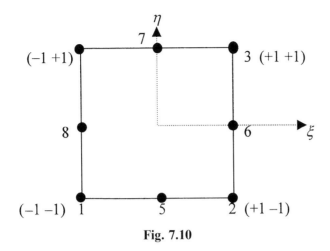

Fig. 7.10

The chosen second order interpolation function that defines the variation of, for example, the elastic displacement ϕ over the element is

$$\phi = a_1 + a_2\,\xi + a_3\,\eta + a_4\,\xi\eta + a_5\,\xi^2 + a_6\,\eta^2 + a_7\,\xi^2\eta + a_8\,\xi\eta^2 \qquad (7.17)$$

Note that the number of terms in this relation and the number of a coefficients is equal to the number of nodes in the element. It follows from (7.17), although we do not lay out the details, that using similar considerations that gave (7.14) from (7.9)

$$\phi = N_1\phi_1 + N_2\phi_2 + N_3\phi_3 + N_4\phi_4 + N_5\phi_5 + N_6\phi_6 + N_7\phi_7 + N_8\phi_8 \qquad (7.18)$$

where the ϕ_i are the eight nodal displacements and the N_i are the shape functions which are each functions of ξ and η.

The shape functions turn out to be

$$N_1 = -1/4(1 - \xi)(1 - \eta)(1 + \xi + \eta)$$

$$N_2 = -1/4(1 + \xi)(1 - \eta)(1 - \xi + \eta)$$

$$N_3 = -1/4(1 + \xi)(1 + \eta)(1 - \xi - \eta)$$

$$N_4 = -1/4(1 - \xi)(1 + \eta)(1 + \xi - \eta) \qquad (7.19)$$

$$N_5 = \tfrac{1}{2}(1 - \xi^2)(1 - \eta)$$

$$N_6 = \tfrac{1}{2}(1 + \xi)(1 - \eta^2)$$

$$N_7 = \tfrac{1}{2}(1 - \xi^2)(1 + \eta)$$

$$N_8 = \tfrac{1}{2}(1 - \xi)(1 - \eta^2)$$

As with the shape functions for the 3-node element, at any node all but one shape function are zero. So for example, at node 1 where the co-ordinates are $(-1-1)$ it is clear from substituting these values in eqn 7.19 that $N_1 = 1$ and N_2 to N_8 are all zero.

We now illustrate the mapping process from the square master element of Fig. 7.10 to a real quadrilateral. One of the valuable features of the parabolic interpolation function that we have chosen is that it is possible to derive real elements with curved sides. Our example does this.

We first set up the nodal co-ordinates of the real element in $x–y$ co-ordinate space, These are shown in Fig. 7.11.

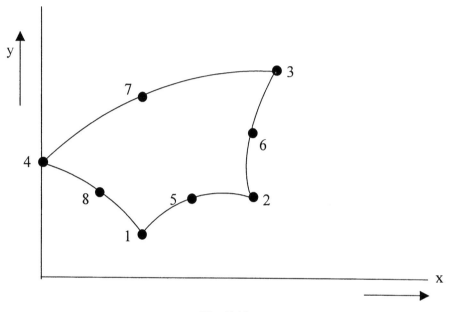

Fig. 7.11

The nodal co-ordinates in Fig. 7.11 are

Node	x	y
1	20	20
2	100	30
3	120	80
4	0	70
5	50	40
6	110	60
7	60	90
8	20	50

The basic mapping relation between the co-ordinates of the master element (ξ, η) in Fig. 7.10 and the co-ordinates of the real element (x, y) in Fig. 7.11 follow the isoparametric form of (7.8) and (7.16) but extended to two dimensions and 8 nodes

$$x = N_1 x_1 + N_2 x_2 + N_3 x_3 + N_4 x_4 + N_5 x_5 + N_6 x_6 + N_7 x_7 + N_8 x_8$$

$$y = N_1 y_1 + N_2 y_2 + N_3 y_3 + N_4 y_4 + N_5 y_5 + N_6 y_6 + N_7 y_7 + N_8 y_8 \qquad (7.20)$$

We will first use these mapping eqns in (7.20) to map points on the master element boundary,

98

that has nodes 1, 5 and 2 (Fig. 7.10) with co-ordinates $(-1, -1)$ $(0, -1)$ and $(+1, -1)$, on to the corresponding boundary carrying nodes 1, 5 and 2 in the real element (Fig. 7.11).

This involves selecting values of the (ξ, η) co-ordinates along the master element boundary and using these to calculate the shape functions from eqn 7.19 and then substituting these and the nodal co-ordinates of the real element (x, y) in eqn 7.20. This gives the mapped (x, y) co-ordinates. We omit the tedious arithmetic and give the final calculated values in the next table.

TABLE 6.2 Relation Between the (ξ, η) Co-ordinates of the Master Element Along the Boundary 1, 5, 2 and the (x, y) Co-ordinates Along the 1, 5, 2 Boundary in the Real Element

ξ	η	x	y
-1	-1	20	20
-0.75	-1	25.625	27.81
-0.5	-1	32.5	33.75
-0.25	-1	40.625	37.81
0	-1	50	40
0.25	-1	60.625	40.31
0.5	-1	72.5	38.75
0.75	-1	85.625	35.31
1	-1	100	30

The first two columns of Table (6.2) give the shape of the master element boundary, and the last two columns give the shape of the corresponding real element boundary. The two profiles are plotted in Figs 7.12a and 7.12b.

1, 5, 2 BOUNDARY IN MASTER ELEMENT

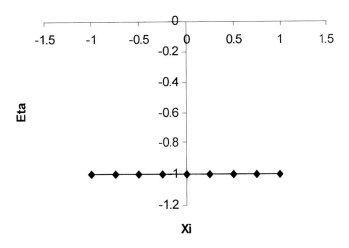

Fig. 7.12a The boundary joining the nodes 1, 5 and 2 in the master element

7.1.3 AN EXAMPLE OF MAPPING A TWO-DIMENSIONAL MESH

The last mapping exercise sets up an orthogonal mesh in the master element and maps it all on to the real element of Fig. 7.11. Each line in the master element mesh is mapped on to the real element in exactly the same way as we mapped the 1,5,2 boundary in the previous exercise. The detailed arithmetic is not presented. The meshes include the 1–5–2 boundaries of Fig. 7.12a and Fig 7.12b. The master element mesh and the mapped real element mesh are graphed in Fig. 7.13a

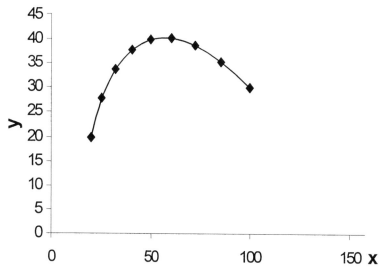

Fig. 7.12b   The mapped 1, 5, 2 boundary in the real element

and b. The juxtaposition of the two meshes emphasises the power of the mapping process.

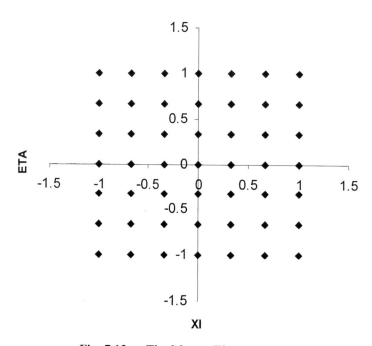

Fig. 7.13a   The Master Element Mesh

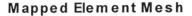

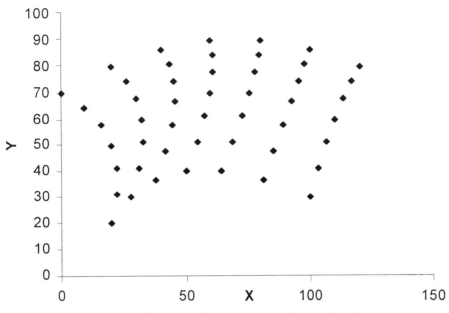

Fig. 7.13b The Real Element Mapped from Fig. 7.13a

7.2 STRESS ANALYSIS USING MAPPED QUADRILATERAL ELEMENTS

In this section we develop the finite element analysis that begins with a square master element and, by mapping using shape functions, transforms this into a quadrilateral of desired shape. We present the analysis using a 4-node element and with shape functions derived from linear interpolation. It is possible in this way to bring out the essential features of the analysis whilst keeping the number of variables and the algebra within manageable bounds. However the linear shape functions that are appropriate for the four-node element yield stresses that are constant across the element. The application of this type of element is seen to best advantage only when it is applied using elements with more nodes and with higher order interpolation functions which give varying stresses over the element. We shall demonstrate this later by working through an example that uses an 8-node element and parabolic interpolation. All the basic features of the theory we shall need to do that example are contained in the 4-node linear analysis which we now develop. At the end of this section we shall show how we can simply extend the 4-node equations to give us the relations we need to work through the 8-node example that follows.

7.2.1 ANALYSIS OF THE FOUR-NODE LINEAR MAPPED QUADRILATERAL ELEMENT

The four-node master element is shown in Fig. 7.14. The position of a point in the element, such as the open circle, is defined by the master co-ordinates (ξ, η). The interpolation relation that describes the variation of elastic displacement over the element and also defines the shape functions, cannot have an order greater than (the number of nodes along the side of the element -1). Since this number is 2 in Fig. 7.14 the order of the interpolation must be 1, that is a linear function must be used. It is

$$\phi = a_1 + a_2\xi + a_3\eta + a_4\xi\eta \qquad (7.21)$$

101

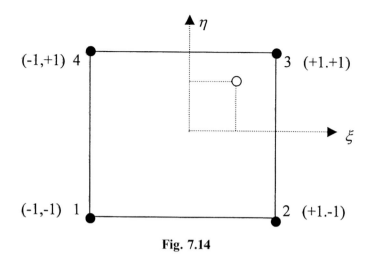

Fig. 7.14

To derive the shape function we follow the same procedure that we did for the linear element. We substitute for each node in turn in equation 7.21 the nodal displacement ϕ_i and the nodal co-ordinates (ξ_i, η_i). This gives four equations which make it possible to eliminate the four unknown coefficients a_1, a_2, a_3 and a_4. By this means we get the standard parametric relation

$$\phi = N_1\phi_1 + N_2\phi_2 + N_3\phi_3 + N_4\phi_4 \tag{7.22}$$

The displacement at any point in the master element in Fig. 7.14 has two components, u and v. Each of these varies with the master element coordinates (ξ, η), which are contained in the shape functions N, according to equation (7.22), that is

$$u = N_1u_1 + N_2u_2 + N_3u_3 + N_4u_4$$

$$v = N_1v_1 + N_2v_2 + N_3v_3 + N_4v_4 \tag{7.23}$$

The shape functions N_i turn out to be

$$N_1 = \tfrac{1}{4}(1 - \xi)(1 - \eta)$$
$$N_2 = \tfrac{1}{4}(1 + \xi)(1 - \eta)$$
$$N_3 = \tfrac{1}{4}(1 + \xi)(1 + \eta) \tag{7.24}$$
$$N_4 = \tfrac{1}{4}(1 - \xi)(1 + \eta)$$

Further, the isoparametric relations that map the co-ordinates of the square master element on to the real quadrilateral element are similar in form to (7.23). They are

$$x = N_1x_1 + N_2x_2 + N_3x_3 + N_4x_4$$

and

$$y = N_1y_1 + N_2y_2 + N_3y_3 + N_4y_4 \tag{7.25}$$

where the N_i are the same shape functions as those that describe the variation of ϕ over the master element. The mapped real element we shall use will look like this

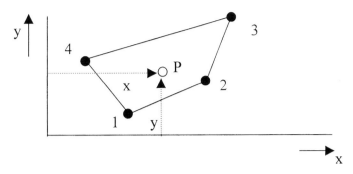

Fig. 7.15

The elastic displacements (u, v) at a point P in the real element with co-ordinates (x, y) are given by exactly the same equations as 7.25 for the master element. The reason for this is important. In eqn (7.23) for the master element, the shape functions contain the master element co-ordinates (ξ, η) so when particular values of these two co-ordinates are substituted, the values of u and v are the elastic displacements at that point. Now the location to which that point is mapped in the real element will carry these values of the displacement with it. So, in summary, the values of (u, v) at the point (x, y) in the real element are the values of (u, v) at the point in the master element (ξ, η) from which (x, y) was mapped. This truth emphasises the generic character of the master serendipity element, which is the root from which all the following analysis flows.

7.2.2 ESSENTIAL DIFFERENCES BETWEEN THIS ANALYSIS AND THAT FOR THE TRIANGULAR CONSTANT STRAIN ELEMENT

It will help to keep the following rather complex analysis in perspective if we explain in what particular it differs from the analysis of the constant strain triangular element of Chapter 6. The essential differences are two:

1. The process of securing the strain matrix, that relates the elastic strains in the element to the nodal displacement, is mathematically more complex and lengthy.
2. We shall wish to analyse elements in which the stress varies across them. This means that the integration through the volume of the element of the strain energy functions and the matrices derived from is no longer trivial, as it was with the constant strain triangular element (see eqns (6.32) and (6.33)). It turns out that numerical methods have to be used, so we need to address them. The integration method we shall choose is called Gauss quadrature.

7.2.3 THE STRAIN MATRIX FOR THE FOUR-NODE ELEMENT

We first develop the matrix that relates the strain at a point in the element to the nodal displacements. The corresponding matrix that we obtained for the constant strain triangular element in Chapter 6 was of the form

$$\{\varepsilon\} = [B]\{u\}. \qquad \text{[eqn (6.13)]}$$

We seek the equivalent of that.

The real quadrilateral element is displayed in Fig. 7.16

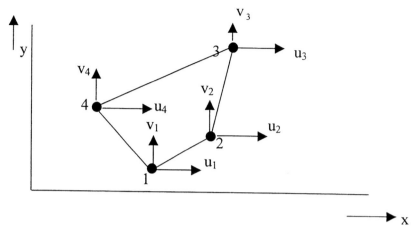

Fig. 7.16

The nodes are numbered 1 to 4 anti-clockwise. The nodal co-ordinates are (x_1, y_1) (x_2, y_2) (x_3, y_3) (x_4, y_4). The nodal displacements in the x and y directions are (u_1, v_1) at node 1 and similarly for the other nodes, so the nodal displacement vector is

$$\{u\} = \{u_1\, v_1\, u_2\, v_2\, u_3\, v_3\, u_4\, v_4\}^T$$

At a point (x, y) in the element the elastic strains are given by the same elastic strain vector that we derived for the triangular element [eqn (6.8)]. We reproduce it here.

$$\{\varepsilon\} = \left\{ \begin{array}{c} \varepsilon_{xx} \\ \varepsilon_{yy} \\ \gamma_{xy} \end{array} \right\} = \left\{ \begin{array}{c} \dfrac{\partial u}{\partial x} \\[2mm] \dfrac{\partial v}{\partial y} \\[2mm] \dfrac{\partial u}{\partial y} + \dfrac{\partial v}{\partial x} \end{array} \right\} \qquad \text{[eqn (6.8)]}$$

We cannot get the partial derivatives with respect to x and y in this elastic strain relation directly because we do not have expressions that relate the displacements u and v directly with x and y. What we do have are the following:

(i) u and v as functions of ξ and η through the isoparametric relations

$$u = N_1 u_1 + N_2 u_2 + N_3 u_3 + N_4 u_4$$
$$v = N_1 v_1 + N_2 v_2 + N_3 v_3 + N_4 v_4. \qquad \text{[eqn 7.23]}$$

We recall that the N terms in these equations are functions of ξ and η, and u_1 to u_4 and v_1 to v_4 are constants for the element, so u and v are functions of ξ and η.

(ii) x and y as functions of ξ and η through the relations

$$x = N_1 x_1 + N_2 x_2 + N_3 x_3 + N_4 x_4$$
$$y = N_1 y_1 + N_2 y_2 + N_3 y_3 + N_4 y_4. \qquad \text{[eqn 7.25]}$$

Again the terms in N are functions of ξ and η and the nodal coordinates x_1 to x_4 and y_1 to y_4 are constants for the element, so x and y are functions of ξ and η.

If we now return to the elastic strain matrix (6.8), the first term we seek is $\frac{\partial u}{\partial x} = \varepsilon_{xx}$. Within the real element, u varies with x and y, but x and y are functions of the coordinates ξ and η in the master element because x and y have been derived by mapping from it. So in mathematical terms, u is a function of x and y which are in turn functions of ξ and η, or

$$u = f[x(\xi, \eta), y(\xi, \eta)]$$

and similar arguments apply to v.

In these circumstances we have to get $\frac{\partial u}{\partial x}$ and the other derivatives in eqn (6.8) by a very roundabout route. Since u and v are functions of another function we have to use the chain rule for differentiation. That is

$$\frac{\partial u}{\partial \xi} = \frac{\partial u}{\partial x}\frac{\partial x}{\partial \xi} + \frac{\partial u}{\partial y}\frac{\partial y}{\partial \xi}$$

and

$$\frac{\partial u}{\partial \eta} = \frac{\partial u}{\partial x}\frac{\partial x}{\partial \eta} + \frac{\partial u}{\partial y}\frac{\partial y}{\partial \eta} \tag{7.26}$$

Casting these in matrix terms

$$\left\{\begin{array}{c} \dfrac{\partial u}{\partial \xi} \\[2ex] \dfrac{\partial u}{\partial \eta} \end{array}\right\} = \begin{bmatrix} \dfrac{\partial x}{\partial \xi} & \dfrac{\partial y}{\partial \xi} \\[2ex] \dfrac{\partial x}{\partial \eta} & \dfrac{\partial y}{\partial \eta} \end{bmatrix} \left\{\begin{array}{c} \dfrac{\partial u}{\partial x} \\[2ex] \dfrac{\partial u}{\partial y} \end{array}\right\} \tag{7.27}$$

The 2×2 matrix in [ ] has an important role in the mapping process and is called a JACOBIAN by mathematicians. It is usually presented with the partial derivatives replaced by equivalent symbols

$$\begin{bmatrix} J_{11} & J_{12} \\ J_{21} & J_{22} \end{bmatrix}$$

So the alternative form of (7.27) is

$$\left\{\begin{array}{c} \dfrac{\partial u}{\partial \xi} \\[2ex] \dfrac{\partial u}{\partial \eta} \end{array}\right\} = \begin{bmatrix} J_{11} & J_{12} \\ J_{21} & J_{22} \end{bmatrix} \left\{\begin{array}{c} \dfrac{\partial u}{\partial x} \\[2ex] \dfrac{\partial u}{\partial y} \end{array}\right\} \tag{7.28}$$

The derivatives $\frac{\partial u}{\partial x}$ and $\frac{\partial u}{\partial y}$ which we need to get the elastic strains from the nodal displacements can now be obtained by inverting eqn (7.28).

$$\left\{\begin{array}{c} \dfrac{\partial u}{\partial x} \\[2ex] \dfrac{\partial u}{\partial y} \end{array}\right\} = \begin{bmatrix} J_{11} & J_{12} \\ J_{21} & J_{22} \end{bmatrix}^{-1} \left\{\begin{array}{c} \dfrac{\partial u}{\partial \xi} \\[2ex] \dfrac{\partial u}{\partial \eta} \end{array}\right\} \tag{7.29}$$

The inversion of the Jacobian matrix is simple. We follow the rules of section 2.4.1. by transposing the matrix of cofactors and dividing by the determinant, det J

$$\det J = [J_{11}J_{22} - J_{21}J_{12}].$$

The cofactors are $\begin{bmatrix} J_{22} & -J_{21} \\ -J_{12} & J_{11} \end{bmatrix}$ and their transpose $\begin{bmatrix} J_{22} & -J_{12} \\ -J_{21} & J_{11} \end{bmatrix}$

So the inverted matrix equation becomes

$$\begin{Bmatrix} \dfrac{\partial u}{\partial x} \\[2mm] \dfrac{\partial u}{\partial y} \end{Bmatrix} = \frac{1}{\det J} \begin{bmatrix} J_{22} & -J_{12} \\ -J_{21} & J_{11} \end{bmatrix} \begin{Bmatrix} \dfrac{\partial u}{\partial \xi} \\[2mm] \dfrac{\partial u}{\partial \eta} \end{Bmatrix} \tag{7.30}$$

A similar set of equations exists for the other displacement component v. Corresponding to (7.26)

$$\frac{\partial v}{\partial \xi} = \frac{\partial v}{\partial x}\frac{\partial x}{\partial \xi} + \frac{\partial v}{\partial y}\frac{\partial y}{\partial \xi}$$

$$\frac{\partial v}{\partial \eta} = \frac{\partial v}{\partial x}\frac{\partial x}{\partial \eta} + \frac{\partial v}{\partial y}\frac{\partial y}{\partial \eta} \tag{7.31}$$

and in matrix form

$$\begin{Bmatrix} \dfrac{\partial v}{\partial \xi} \\[2mm] \dfrac{\partial v}{\partial \eta} \end{Bmatrix} = \begin{bmatrix} \dfrac{\partial x}{\partial \xi} & \dfrac{\partial y}{\partial \xi} \\[2mm] \dfrac{\partial x}{\partial \eta} & \dfrac{\partial y}{\partial \eta} \end{bmatrix} \begin{Bmatrix} \dfrac{\partial v}{\partial x} \\[2mm] \dfrac{\partial v}{\partial y} \end{Bmatrix} \tag{7.32}$$

This matrix equation has the same Jacobian as does eqn (7.27) so the inverse matrix will be similar

$$\begin{Bmatrix} \dfrac{\partial v}{\partial x} \\[2mm] \dfrac{\partial v}{\partial y} \end{Bmatrix} = \frac{1}{\det J} \begin{bmatrix} J_{22} & -J_{12} \\ -J_{21} & J_{11} \end{bmatrix} \begin{Bmatrix} \dfrac{\partial v}{\partial \xi} \\[2mm] \dfrac{\partial v}{\partial \eta} \end{Bmatrix} \tag{7.33}$$

Equations (7.30) and (7.33) contain all the partial derivatives that we need to evaluate the three elastic strains. It is tidy to collect them all together in a single matrix.

$$\begin{Bmatrix} \dfrac{\partial u}{\partial x} \\[2mm] \dfrac{\partial u}{\partial y} \\[2mm] \dfrac{\partial v}{\partial x} \\[2mm] \dfrac{\partial v}{\partial y} \end{Bmatrix} = \frac{1}{\det J} \begin{bmatrix} J_{22} & -J_{12} & 0 & 0 \\ -J_{21} & J_{11} & 0 & 0 \\ 0 & 0 & J_{22} & -J_{12} \\ 0 & 0 & -J_{21} & J_{11} \end{bmatrix} \begin{Bmatrix} \dfrac{\partial u}{\partial \xi} \\[2mm] \dfrac{\partial u}{\partial \eta} \\[2mm] \dfrac{\partial v}{\partial \xi} \\[2mm] \dfrac{\partial v}{\partial \eta} \end{Bmatrix} \tag{7.34}$$

All the terms on the right of the eqn (7.34) are known and so therefore are the left-hand side derivatives. These are the ones we seek for the strain matrix, which we are now in a position to evaluate.

106

Returning therefore to the strain matrix

$$
\left\{ \begin{array}{c} \varepsilon_{xx} \\ \varepsilon_{yy} \\ \gamma_{xy} \end{array} \right\} = \left\{ \begin{array}{c} \dfrac{\partial u}{\partial x} \\[2mm] \dfrac{\partial v}{\partial y} \\[2mm] \dfrac{\partial u}{\partial y} + \dfrac{\partial v}{\partial x} \end{array} \right\}
$$

For ε_{xx} we multiply out the first row of (7.34), which is the equation for $\frac{\partial u}{\partial x}$

$$
\frac{\partial u}{\partial x} = \frac{1}{\det J} \left[J_{22} \frac{\partial u}{\partial \xi} - J_{12} \frac{\partial u}{\partial \eta} \right] \tag{7.35}
$$

For ε_{yy} we multiply out the fourth row of (7.34), which is the equation for $\frac{\partial v}{\partial y}$

$$
\frac{\partial v}{\partial y} = \frac{1}{\det J} \left[- J_{21} \frac{\partial v}{\partial \xi} + J_{11} \frac{\partial v}{\partial \eta} \right] \tag{7.36}
$$

For γ_{xy} we multiply out the second and third rows, which are the equations respectively for $\frac{\partial u}{\partial y}$ and $\frac{\partial v}{\partial x}$, so

$$
\frac{\partial u}{\partial y} + \frac{\partial v}{\partial x} = \frac{1}{\det J} \left[- J_{21} \frac{\partial u}{\partial \xi} + J_{11} \frac{\partial u}{\partial \eta} + J_{22} \frac{\partial v}{\partial \xi} - J_{12} \frac{\partial v}{\partial \mu} \right] \tag{7.37}
$$

We collect (7.35), (7.36) and (7.37) together in the strain matrix

$$
\varepsilon = \left\{ \begin{array}{c} \dfrac{\partial u}{\partial x} \\[2mm] \dfrac{\partial v}{\partial y} \\[2mm] \dfrac{\partial u}{\partial y} + \dfrac{\partial v}{\partial x} \end{array} \right\} = \frac{1}{\det J} \begin{bmatrix} J_{22} & -J_{12} & 0 & 0 \\ 0 & 0 & -J_{21} & J_{11} \\ -J_{21} & J_{11} & J_{22} & -J_{12} \end{bmatrix} \left\{ \begin{array}{c} \dfrac{\partial u}{\partial \xi} \\[2mm] \dfrac{\partial u}{\partial \eta} \\[2mm] \dfrac{\partial v}{\partial \xi} \\[2mm] \dfrac{\partial v}{\partial \eta} \end{array} \right\} \tag{7.38}
$$

The terms in this matrix equation contain only

1. The derivatives of the shape functions, which are known,
2. The nodal coordinates of the real element, which are known,
3. The nodal displacements, which are the principal unknowns that we seek.

So equation (7.38) is the relation which we set out to establish at the beginning of this Section 7.2.3 between the strain at a point in the element and the nodal displacements.

We need finally to go through some tedious manipulation to make (7.38) more explicit and get it in the form that we shall use for computation. First we look at the vector of derivatives

$$
\left\{ \frac{\partial u}{\partial \xi}, \frac{\partial u}{\partial \eta}, \frac{\partial v}{\partial \xi}, \frac{\partial v}{\partial \eta} \right\}^T
$$

$$
u = N_1 u_1 + N_2 u_2 + N_3 u_3 + N_4 u_4 \qquad \text{[equation (7.23)]}
$$

107

so

$$\frac{\partial u}{\partial \xi} = \frac{\partial N_1}{\partial \xi} u_1 + \frac{\partial N_2}{\partial \xi} u_2 + \frac{\partial N_3}{\partial \xi} u_3 + \frac{\partial N_4}{\partial \xi} u_4$$

because the u_i are constants

also

$$\frac{\partial u}{\partial \eta} = \frac{\partial N_1}{\partial \eta} v_1 + \frac{\partial N_2}{\partial \eta} v_2 + \frac{\partial N_3}{\partial \eta} v_3 + \frac{\partial N_4}{\partial \eta} v_4$$

because the v_i are constants

and

$$v = N_1 v_1 + N_2 v_2 + N_3 v_3 + N_4 v_4 \qquad \text{[eqn (7.23)]}$$

so

$$\frac{\partial v}{\partial \xi} = \frac{\partial N_1}{\partial \xi} v_1 + \frac{\partial N_2}{\partial \xi} v_2 + \frac{\partial N_3}{\partial \xi} v_3 + \frac{\partial N_4}{\partial \xi} v_4$$

also

$$\frac{\partial v}{\partial \eta} = \frac{\partial N_1}{\partial \eta} v_1 + \frac{\partial N_2}{\partial \eta} v_2 + \frac{\partial N_3}{\partial \eta} v_3 + \frac{\partial N_4}{\partial \eta} v_4$$

We gather these four equations together in a matrix

$$\begin{Bmatrix} \dfrac{\partial u}{\partial \xi} \\ \dfrac{\partial u}{\partial \eta} \\ \dfrac{\partial v}{\partial \xi} \\ \dfrac{\partial v}{\partial \eta} \end{Bmatrix} = \begin{bmatrix} \dfrac{\partial N_1}{\partial \xi} & 0 & \dfrac{\partial N_2}{\partial \xi} & 0 & \dfrac{\partial N_3}{\partial \xi} & 0 & \dfrac{\partial N_4}{\partial \xi} & 0 \\ \dfrac{\partial N_1}{\partial \eta} & 0 & \dfrac{\partial N_2}{\partial \eta} & 0 & \dfrac{\partial N_3}{\partial \eta} & 0 & \dfrac{\partial N_4}{\partial \eta} & 0 \\ 0 & \dfrac{\partial N_1}{\partial \xi} & 0 & \dfrac{\partial N_2}{\partial \xi} & 0 & \dfrac{\partial N_3}{\partial \xi} & 0 & \dfrac{\partial N_4}{\partial \xi} \\ 0 & \dfrac{\partial N_1}{\partial \eta} & 0 & \dfrac{\partial N_2}{\partial \eta} & 0 & \dfrac{\partial N_3}{\partial \eta} & 0 & \dfrac{\partial N_4}{\partial \eta} \end{bmatrix} \begin{Bmatrix} u_1 \\ v_1 \\ u_2 \\ v_2 \\ u_3 \\ v_3 \\ u_4 \\ v_4 \end{Bmatrix} \quad (7.39)$$

The right-hand side of (7.39) will be given the condensed form

$$= [G]\{u\} \qquad (7.40)$$

We can derive explicit expressions for the derivatives of the shape functions in $[G]$. Recalling (7.24)

$N_1 = \frac{1}{4}(1-\xi)(1-\eta)$ so $\dfrac{\partial N_1}{\partial \xi} = -\frac{1}{4}(1-\eta)$ $\dfrac{\partial N_1}{\partial \eta} = -\frac{1}{4}(1-\xi)$

$N_2 = \frac{1}{4}(1+\xi)(1-\eta)$ so $\dfrac{\partial N_2}{\partial \xi} = \frac{1}{4}(1-\eta)$ $\dfrac{\partial N_2}{\partial \eta} = -\frac{1}{4}(1+\xi)$

$N_3 = \frac{1}{4}(1+\xi)(1+\eta)$ so $\dfrac{\partial N_3}{\partial \xi} = \frac{1}{4}(1+\eta)$ $\dfrac{\partial N_2}{\partial \eta} = -\frac{1}{4}(1+\xi)$

$N_4 = \frac{1}{4}(1-\xi)(1+\eta)$ so $\dfrac{\partial N_4}{\partial \xi} = -\frac{1}{4}(1+\eta)$ $\dfrac{\partial N_4}{\partial \eta} = \frac{1}{4}(1-\xi)$ (7.41)

Let us make a brief aside to remind ourselves about partial differentiation. To differentiate N_1 partially with respect to ξ we treat η as a constant so

$$\frac{\partial}{\partial \xi}\left[\frac{1}{4}(1-\xi)(1-\eta)\right] = \frac{1}{4}(1-\eta)\frac{\partial}{\partial \xi}(1-\xi) = -\frac{1}{4}(1-\eta)$$

similarly if we differentiate N_1 partially with respect to η we regard ξ as constant so

$$\frac{\partial}{\partial \eta}\left[\frac{1}{4}(1-\xi)(1-\eta)\right] = \frac{1}{4}(1-\xi)\frac{\partial}{\partial \eta}(1-\eta) = -\frac{1}{4}(1-\xi)$$

If we substitute the expressions for the shape function derivatives in terms of ξ and η in the matrix for $[G]$,

$$[G] = \frac{1}{4}\begin{bmatrix} -(1-\eta) & 0 & (1-\eta) & 0 & (1+\eta) & 0 & -(1+\eta) & 0 \\ -(1-\xi) & 0 & -(1+\xi) & 0 & (1+\xi) & 0 & (1-\xi) & 0 \\ 0 & -(1-\eta) & 0 & (1-\eta) & 0 & (1+\eta) & 0 & -(1+\eta) \\ 0 & -(1-\xi) & 0 & -(1+\xi) & 0 & (1+\xi) & 0 & (1-\xi) \end{bmatrix} \quad (7.42)$$

Finally we express the terms from the Jacobian matrix in (7.38) in terms of ξ and η.

$$J_{11} = \frac{\partial x}{\partial \xi} \quad \text{but } x = N_1 x_1 + N_2 x_2 + N_3 x_3 + N_4 x_4$$

So

$$\frac{\partial x}{\partial \xi} = \frac{\partial N_1}{\partial \xi}x_1 + \frac{\partial N_2}{\partial \xi}x_2 + \frac{\partial N_3}{\partial \xi}x_3 + \frac{\partial N_4}{\partial \xi}x_4$$

Substituting for the derivatives from (7.41)

$$J_{11} = \frac{\partial x}{\partial \xi} = \frac{1}{4}[-(1-\eta)x_1 + (1-\eta)x_2 + (1+\eta)x_3 - (1+\eta)x_4] \quad (7.43)$$

Similarly we find

$$J_{12} = \frac{\partial y}{\partial \xi} = \frac{1}{4}[-(1-\eta)y_1 + (1-\eta)y_2 + (1+\eta)y_3(1+\eta)y_4] \quad (7.44)$$

$$J_{21} = \frac{\partial x}{\partial \xi} = \frac{1}{4}[-(1-\xi)x_1 - (1+\xi)x_2 + (1+\xi)x_3 + (1-\xi)x_4] \quad (7.45)$$

$$J_{22} = \frac{\partial y}{\partial \eta} = \frac{1}{4}[-(1-\xi)y_1 - (1+\xi)y_2 + (1+\xi)y_3 + (1-\xi)y_4] \quad (7.46)$$

We can now return to the elastic strain matrix equation (7.38) and express it as

$$\{\varepsilon\} = \left\{ \begin{array}{c} \dfrac{\partial u}{\partial x} \\[2mm] \dfrac{\partial v}{\partial y} \\[2mm] \dfrac{\partial u}{\partial y} + \dfrac{\partial v}{\partial x} \end{array} \right\} = \frac{1}{\det J} \begin{bmatrix} J_{22} & -J_{12} & 0 & 0 \\ 0 & 0 & -J_{21} & J_{11} \\ -J_{21} & J_{11} & J_{22} & -J_{12} \end{bmatrix} [G]\{u\} \qquad (7.47)$$

where we have replaced the vector of partial derivatives

$$\left\{ \begin{array}{c} \dfrac{\partial u}{\partial \xi} \\[2mm] \dfrac{\partial u}{\partial \eta} \\[2mm] \dfrac{\partial v}{\partial \xi} \\[2mm] \dfrac{\partial v}{\partial \eta} \end{array} \right\} \quad \text{with its abbreviated form (7.40) } [G]\{u\}.$$

If in (7.47) we let

$$[A] = \frac{1}{\det J} \begin{bmatrix} J_{22} & -J_{12} & 0 & 0 \\ 0 & 0 & -J_{21} & J_{11} \\ -J_{21} & J_{11} & J_{22} & -J_{12} \end{bmatrix} \qquad (7.48)$$

then we can express (7.47) in the condensed form

$$\{\varepsilon\} = [A][G]\{u\}$$

$$\text{or} \qquad \{\varepsilon\} = [B]\{u\} \quad \text{where} \quad [B] = [A][G] \qquad (7.49)$$

This relates the elastic strains to the nodal displacements and is the relation we set out to establish at the beginning of this section. Equation (7.49) is the equivalent of eqn (6.13) for the constant strain triangular element, but (6.13) was arrived at by a very much simpler process than the protracted and rather tedious manipulations that were necessary to achieve (7.49).

7.2.4 EXTENSION OF THE FOUR-NODE RELATIONS TO ESTABLISH THE STRAIN MATRIX FOR THE EIGHT-NODE ELEMENT

In the examples that follow this analysis, we shall use an 8-node quadrilateral element. The attraction of this choice is that it makes it possible to use parabolic shape functions. This in turn yields elastic strains and stresses that vary over the element, in contrast to the 4-node quadrilateral and the 3-node triangle in which the strains and stresses are constant.

We have laid out the basic analysis that leads to the strain matrix [B] using the 4-node element in order to keep the number of terms in the various relations to a minimum. The corresponding expressions for the 8-node element have a structure similar to those for the 4-node element, the only difference being that there are more, but similar, terms. We now lay out those differences and in the process list the relations we shall need for the 8-node examples that follow.

Our objective is to secure the expressions for the Strain matrix [B]. This, we recall, relates the strain vector $\{\varepsilon\}$ to the nodal displacement vector u through

$$\varepsilon = [B]\{u\}.$$

In the 8-node case as with the 4-node element [B] is the product of the two matrices

$$[B] = [A][G]$$

The 8-node [A] matrix contains exactly the same array of Jacobians, that is

$$[A] = \frac{1}{\det J} \begin{bmatrix} J_{22} & J_{12} & 0 & 0 \\ 0 & 0 & J_{21} & J_{22} \\ J_{21} & J_{11} & J_{22} & J_{12} \end{bmatrix}$$

However since there are eight shape functions associated with the eight nodes, which are listed in eqn (7.19) of Section 7.1.2, each Jacobian has eight terms. We quote these below

$$J_{11} = s1.x1 + s2.x2 + s3.x3 + s4.x4 + s5.x5 + s6.x6 + s7.x7 + s8.x8$$

$$J_{12} = s1.y1 + s2.y2 + s3.y3 + s4.y4 + s5.y5 + s6.y6 + s7.y7 + s8.y8$$

$$J_{21} = t1.x1 + t2.x2 + t3.x3 + t4.x4 + t5.x5 + t6.x6 + t7.x7 + t8.x8$$

$$J_{22} = t1.y1 + t2.y2 + t3.y3 + t4.y4 + t5.y5 + t6.y6 + t7.y7 + t8.y8$$

Where

$$s1 = \frac{\partial N_1}{\partial \xi} \quad s2 = \frac{\partial N_2}{\partial \xi} \quad s3 = \frac{\partial N_3}{\partial \xi}$$

and so on

$$t1 = \frac{\partial N_1}{\partial \eta} \quad t2 = \frac{\partial N_2}{\partial \eta} \quad t3 = \frac{\partial N_3}{\partial \eta}$$

and so on.

$x1$ to $x8$ are the nodal x-coordinates and $y1$ to $y8$ are the nodal y-coordinates. $\det J = J_{11}J_{22} - J_{21}J_{12}$ as before

The 8-node [G] matrix is

$$[G] = \begin{bmatrix} s1 & 0 & s2 & 0 & s3 & 0 & s4 & 0 & s5 & 0 & s6 & 0 & s7 & 0 & s8 & 0 \\ t1 & 0 & t2 & 0 & t3 & 0 & t4 & 0 & t5 & 0 & t6 & 0 & t7 & 0 & t8 & 0 \\ 0 & s1 & 0 & s2 & 0 & s3 & 0 & s4 & 0 & s5 & 0 & s6 & 0 & s7 & 0 & s8 \\ 0 & t1 & 0 & t2 & 0 & t3 & 0 & t4 & 0 & t5 & 0 & t6 & 0 & t7 & 0 & t8 \end{bmatrix}$$

where the s and t terms have been defined above. This $[G]$ matrix is similar to the 4-node matrix but with twice as many terms.

Finally, for reference, we list the 16 derivatives of the 8-node shape functions. The shape functions are given in eqn (7.19) of Section 7.1.2.

We remind ourselves again about the process of partial differentiation by working out the ξ derivative of the first shape function

$$N_1 = -0.25(1 - \xi)(1 - \eta)(1 + \xi + \eta)$$

$$\frac{\partial N_1}{\partial \xi} = -0.25(1 - \eta)\frac{\partial}{\partial \xi}(1 - \xi)(1 + \xi + \eta)$$

$$= -0.25(1 - \eta)\{(1 - \xi)\frac{\partial}{\partial \xi}(1 + \xi + \eta) + (1 + \xi + \eta)\frac{\partial}{\partial \xi}(1 - \xi)\}$$

$$= -0.25(1 - \eta)\{1 x(1 - \xi) + (-1)(1 + \xi + \eta)\}$$

$$= 0.25(1 - \eta)(2\xi - \eta)$$

The complete list of derivatives is

$$s1 = \frac{\partial N_1}{\partial \xi} = 0.25(1 - \eta)(2\xi + \eta) \qquad\qquad t1 = \frac{\partial N_1}{\partial \eta} = 0.25(1 - \xi)(2\eta + \xi)$$

$$s2 = \frac{\partial N_2}{\partial \xi} = 0.25(1 - \eta)(2\xi - \eta) \qquad\qquad t2 = \frac{\partial N_2}{\partial \eta} = -0.25(1 + \xi)(\xi - 2\eta)$$

$$s3 = \frac{\partial N_3}{\partial \xi} = 0.25(1 + \eta)(2\xi + \eta) \qquad\qquad t3 = \frac{\partial N_3}{\partial \eta} = 0.25(1 + \eta)(2\eta + \xi)$$

$$s4 = \frac{\partial N_4}{\partial \xi} = 0.25(1 + \eta)(2\xi - \eta) \qquad\qquad t4 = \frac{\partial N_4}{\partial \eta} = 0.25(1 - \xi)(2\eta - \xi)$$

$$s5 = \frac{\partial N_5}{\partial \xi} = -\xi(1 - \eta) \qquad\qquad t5 = \frac{\partial N_5}{\partial \eta} = -0.5(1 - \xi^2)$$

$$s6 = \frac{\partial N_6}{\partial \xi} = 0.5(1 - \eta^2) \qquad\qquad t6 = \frac{\partial N_6}{\partial \eta} = -\eta(1 + \xi)$$

$$s7 = \frac{\partial N_7}{\partial \xi} = -\xi(1 + \eta) \qquad\qquad t7 = \frac{\partial N_7}{\partial \eta} = 0.5(1 - \xi^2)$$

$$s8 = \frac{\partial N_8}{\partial \xi} = -0.5(1 - \eta^2) \qquad\qquad t8 = \frac{\partial N_8}{\partial \eta} = -\eta(1 - \xi)$$

We are now in a position to set up the stiffness matrix for the quadrilateral element.

7.3   THE STIFFNESS MATRIX OF THE QUADRILATERAL ELEMENT

With the relation between the nodal displacements and the elastic strains [eqn (7.49)], established for the quadrilateral element the way forward to the stiffness matrix follows the same basic analysis that we developed for the triangular element. The important difference is the following:

In the triangular element analysis we developed an expression [eqn (6.32)] for the elastic strain energy generated by the virtual displacements. This involved an integral through the volume of the element. Since the integrand was constant throughout the element, the integral was easily worked out as the integrand times the element volume. We now wish to analyse a situation with the quadrilateral element where the stress, and therefore the integrand in (6.32), varies across the element so the integration is more complex. We must retain the integral sign in (6.32).

If we pursue the analysis in Chapter 6 from equation (6.32) and retain the integral sign, we reach eqn (6.48) which now has the form

$$\int_V [B]^T \{\sigma\} \mathrm{d}V = \{F\} \qquad (7.50)$$

and

$$\{\sigma\} = [D][B]\{u\}$$

so

$$\int_V [B]^T [D][B]\{u\}.\mathrm{d}V = \{F\}$$

But the nodal displacements in the vector $\{u\}$ are constant so we can take it outside the integral sign then

$$\left[\int_V [B]^T [D][B].\mathrm{d}V \right] \{u\} = \{F\}.$$

This is of the form

$$[K]\{u\} = \{F\}$$

where $[K]$ is the stiffness matrix and so

$$[K] = \left[\int_V [B]^T [D][B].\mathrm{d}V \right] \qquad (7.51)$$

To establish the stiffness matrix we need to perform the integral of (7.51) through the volume of the element and the function to be integrated is $[B]^T [D][B]$. The product of the three matrices in this integrand yields a matrix of order $[8 \times 8]$ for the quadrilateral element. The integration of a matrix is achieved by integrating each term. In this case there are 64 terms. We shall see how this integration is achieved in the next section.

7.3.1   THE INTEGRATION NEEDED TO FORM THE STIFFNESS MATRIX

If in the stiffness matrix integral (7.51) we let the small volume $\mathrm{d}V$ in the real element have an area $\mathrm{d}x.\mathrm{d}y$ in a sheet of uniform thickness t then

$$\mathrm{d}V = t\mathrm{d}x.\mathrm{d}y \qquad \text{so the stiffness matrix integral is}$$

$$[K] = t \iint [B]^T [D][B] \, \mathrm{d}x.\mathrm{d}y \qquad (7.52)$$

The evaluation of this integral is complicated by the fact that the matrices in the integrand are expressed in terms of ξ and η not x and y. The way forward is to perform the integral in the master element, where the coordinates are ξ and η, and we use the mapping process to relate the element dx.dy in the real element to the corresponding element dξ.dη in the master element, as the following diagram shows.

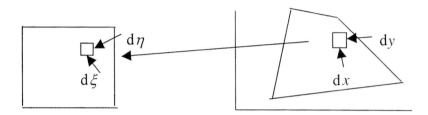

It turns out that the area of the real calculus volume element is

$$\mathrm{d}x.\mathrm{d}y = \det J \, \mathrm{d}\xi.\mathrm{d}\eta \qquad (7.53)$$

we justify and prove this relation in Appendix (7.1)

We recall that $\det J$ is the determinant of the Jacobian, which is

$$\det J = [J_{11}J_{22} - J_{21}J_{12}]$$

Physically $\det J$ is the ratio of the area of the real calculus element to the area of the master calculus element. Since the thickness t of the sheet is the same for both $\det J$ is also the ratio of volumes. We change the variables in (7.52) by substituting (7.53)

$$[K] = t \iint [B]^T[D][B] \det J \, \mathrm{d}\xi \mathrm{d}\eta \qquad (7.54)$$

The advantage of this transformation from x and y to ξ and η is that the integration is performed over the same element, the simple square master element with sides of 2 units, irrespective of the size and shape of the real element.

The final problem in the development of the analysis is that (7.54) cannot be integrated directly so numerical integration has to be used. We now explain how that is achieved.

7.3.2 NUMERICAL INTEGRATION IN ONE DIMENSION BY GAUSS QUADRATURE

The precise value of the integral of a function between prescribed limits of the independent variable is the area under the curve representing the graph of the function between these limits. Fig. 7.17 illustrates this. The

$$\int_{-1}^{+1} f(x)\mathrm{d}x$$

is the total area under the graph between $x = -1$ and $x = +1$. This is the sum of the area of all the strips like that with dotted sides in Fig. 7.17, which has an area $f(x).\mathrm{d}x$, between $x = -1$ and $x = +1$.

A popular numerical integration method for finite element analysis is Gaussian quadrature. This involves approximating the area under the curve representing the function by a series of

114

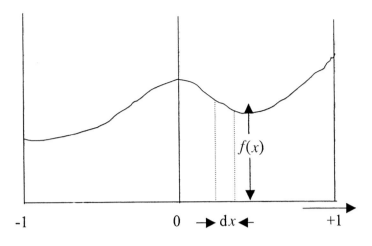

Fig. 7.17

rectangles. The height of each rectangle is the value of the function at what is called the sampling point, and the width is called the weight. For integrals that are our concern, with limits +1 and −1, the sampling points are symmetrical about the origin. For example if 3 sample points are chosen the best approximation to the integral is achieved if the three points are at

$$x = -0.77460 \qquad x = 0 \qquad \text{and} \qquad x = +0.77640$$

and the weights, which are the widths of the rectangles are 5/9, 8/9 and 5/9. The determination of the integral by this method is illustrated in Fig. 7.18. We identify the three sampling points 1, 2 and 3 in Fig (7.18) The value of

$$\int_{-1}^{+1} f(x).dx$$

is the sum of the areas of the three rectangles A, B and C. Rectangle A has a width (the WEIGHT) of 5/9 and the height is the value of the function $f(x)$ at $x = x_1 = -0.77460$ (the first SAMPLING POINT) and similarly for the areas B and C.

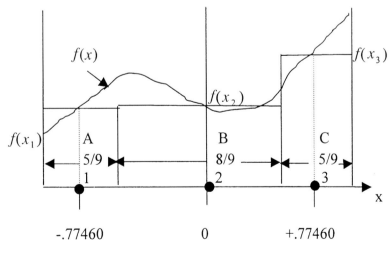

Fig. 7.18

115

BASIC PRINCIPLES OF THE FINITE ELEMENT METHOD

The formal statement of the relation for the Gauss quadrature integration with three sampling points is

$$\int_{-1}^{+1} f_1(x)dx = w_1 f_1(x) + w_2 f_2(x) + w_3 f_3(x) \tag{7.55}$$

where w_1 is the weight at sampling point 1 and $f_1(x)$ is the value of the function to be integrated at sampling point 1 and correspondingly for sampling points 2 and 3.

It is helpful to illustrate this method by evaluating a particular integral. Let us work out

$$\int_{-1}^{+1} \cos(\pi x/2).dx$$

There is of course an exact solution to this, it is

$$\left[\frac{2}{\pi}\sin\left(\frac{\pi x}{2}\right)\right]_{-1}^{+1} = \frac{4}{\pi} = 1.2732$$

Now let us get a value by 3-point Gauss Quadrature. The sampling points are at $x = -0.77460$ $x = 0$ and $x = +0.77460$.

At $x = -0.77460$, $f(x) = \cos(\pi x/2) = 0.34671$ and the weight is $w_A = 5/9$.

At $x = 0$ $f(x)$ $= 1.000$ and the weight is $w_B = 8/9$.

At $x = +0.77460$ $f(x)$ $= 0.36471$ and the weight is $w_C = 5/9$.

So the Gauss Quadrature value of the integral is

$$(0.34671 \times 5/9) + (1.000 \times 8/9) + (0.36471 \times 5/9) = 1.2741$$

which is very close to the exact value 1.2732.

Gauss quadrature gives an exact value for the integral of a polynomial of degree $(2n - 1)$ if n sampling points are used. So a cubic polynomial of degree 3 will be exactly integrated with 2 sample points because $(2 \times 2 - 1) = 3$. A fifth order polynomial will be exactly integrated with 3 sampling points $-(2 \times 3 - 1) = 5$.

The location of the sampling points and the weight attached to each point is given in the following table for 1, 2, 3 and 4 sampling points.

Number of Sampling points	x-coordinate of sampling point	weight w
1	0	2.0
2	−0.57735	1.0
	+0.57735	1.0
3	−0.774597	5/9
	0	8/9
	+0.774597	5/9
4	−0.861136	0.347855
	−0.339981	0.652145
	+0.339981	0.652145
	+0.861136	0.347855

We can see how the sample points and weights are arrived at in the following analysis for two sample points.

The Gauss Quadrature relation for the numerical integration with two sampling point is

$$\int_{-1}^{+1} f(x).\mathrm{d}x = w_1 f(x_1) + w_2 f(x_2)$$

w_1 is the weight at the first sampling point $x = x_1$ where the value of $f(x)$ is $f(x_1)$ and w_2 and $f(x_2)$ refer to the second sampling point at $x = x_2$.

Assume that the function is a cubic polynomial

$$f(x) = a_0 + a_1 x + a_2 x^2 + a_3 x^3 \qquad (7.55)$$

The error in the Gauss quadrature value is the difference between the exact answer and the Gauss quadrature estimate i.e.

$$\int_{-1}^{+1} (a_0 + a_1 x + a_2 x^2 + a_3 x^3)\mathrm{d}x - [w_1 f(x_1) + w_2 f(x_2)]$$

If the Gauss quadrature value is to be exact this error will be zero. We seek the condition for this to be so. The value of the integral is

$$\left[a_0 x + a_1 \frac{x^2}{2} + a_2 \frac{x^3}{3} + a_3 \frac{x^4}{4} \right]_{-1}^{+1} = 2a_0 + \frac{2a_2}{3}$$

So the zero error condition is

$$\left[2a_0 + \frac{2a_2}{3} \right] - [w_1 f(x_1) + w_2 f(x_2)] = 0$$

Expanding $f(x_1)$ and $f(x_2)$ with (7.55)

$$\left[2a_0 + \frac{2a_2}{3} \right] - [w_1(a_0 + a_1 x_1 + a_2 x_1^2 + a_3 x_1^3)] - [w_2(a_0 + a_1 x_2 + a_2 x_2^2 + a_3 x_2^3)] = 0$$

Reordering the terms in this equation

$$a_0[2 - (w_1 + w_2)] - a_1[w_1 x_1 + w_2 x_2] + a_2\left[\frac{2}{3} - (w_1 x_1^2 + w_2 x_2^2) \right] - a_3[w_1 x_1^3 + w_2 x_2^3] = 0$$

For the left hand side to be zero, each term must be zero so

$$2 - (w_1 + w_2) = 0 \quad \text{so} \quad w_1 + w_2 = 2 \qquad (7.56)$$

$$w_1 x_1 + w_2 x_2 = 0 \qquad (7.57)$$

but the sampling points are symmetrical about the origin so

$$x_1 = -x_2 \qquad (7.58)$$

Substituting (7.58) in (7.57)

$$x_1(w_1 - w_2) = 0 \quad \text{therefore} \quad (w_1 - w_2) = 0 \quad \text{and} \quad w_1 = w_2$$

Substituting this in 7.56

$$2w_1 = 2 \quad \text{or} \quad w_1 = 1 \quad \text{and} \quad w_2 = 1$$

And putting

$$w_1 = w_2 = 1 \text{ in } (7.57) \quad x_1 + x_2 = 0 \quad \text{so} \quad x_1 = -x_2$$

So both weights are unity.

Putting the coefficient of a_2 equal to zero

$$\frac{2}{3} - (w_1 x_1^2 + w_2 x_2^2) = 0 \quad \text{but} \quad w_1 = w_2 = 1 \quad \text{and} \quad x_1 = -x_2$$

$$\text{so} \quad \frac{2}{3} - 2w_1 x_1^2 = 0 \quad \text{or} \quad w_1 x_1^2 = \frac{1}{3} \quad \text{and} \quad w_1 = 1$$

$$\text{so} \quad x_1 = \left[\frac{1}{3}\right]^{1/2} = 0.57735 \quad \text{and} \quad x_2 = -x_1 = -0.57735$$

We conclude therefore that the sampling points are at

$$x = 0.57735 \quad \text{and} \quad x = -0.57735 \quad \text{and the weight of both points is } w = 1.]$$

These conclusions are illustrated in the following diagram.

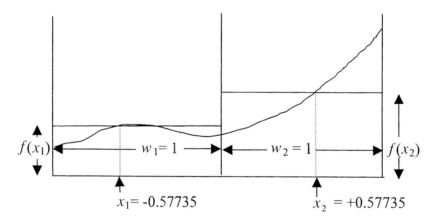

Fig. 7.19 Diagram illustrating the integration of a function using two Gauss points

7.3.3 Two-dimensional Integration by Gauss Quadrature

In finite element analysis, we need frequently to integrate functions over the whole area of an element. This involves double integration with respect to two variables. We will now illustrate the method by performing an integration over the square serendipity element. The two variables will be the serendipity element co-ordinates ξ and η. We shall use the Gauss quadrature method with, in the first instance, four integration points.

We saw in Section (7.2.5) eqn (7.54) that, in the process of determining the stiffness matrix, we need to evaluate the following integral:

$$\int\limits_{-1}^{+1} \int\limits_{-1}^{+1} f(\xi, \eta).d\xi.d\eta \qquad (7.59)$$

where $f(\xi, \eta)$ is a complex function of ξ and η, the details of which need not concern us for our present purpose. Of course if we wish to evaluate a particular integral, as we will later, the details of the function will be required.

Let us look at the serendipity calculus element which is drawn in Fig. 7.20.

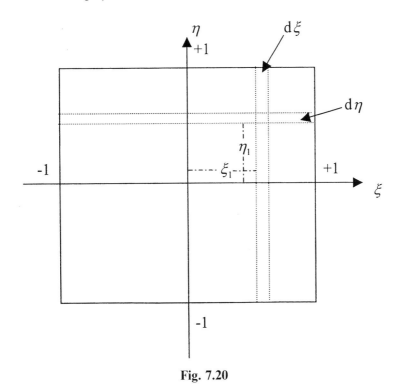

Fig. 7.20

We shall need to distinguish between those elements that are finite elements and those that are used in the calculus. To avoid confusion we shall call the latter 'calculus elements'. The small calculus element in Fig. 7.20 at the point (ξ_1, η_1) has an area $d\xi.d\eta$ and the value at that point of the function to be integrated is $f(\xi_1, \eta_1)$, that is the value of the function when ξ_1 is substituted for ξ and η_1 is substituted for η. Now

$$f(\xi_1, \eta_1).d\xi d\eta \qquad (7.60)$$

is the contribution to the integral from the value of the function, assumed constant, over the calculus element $d\xi d\eta$.

To evaluate the complete integral we need to sum (7.60) over all the calculus elements of the serendipity square. We do this in two steps. First we perform the sum over the strip of length 2, extending from $\xi = -1$ to $\xi = +1$ and of width $d\eta$ with η constant. This strip is drawn in Fig. 7.21. The contribution from this strip to the overall integral is

$$\left[\int\limits_{-1}^{+1} f(\xi, \eta).d\xi \right].d\eta \qquad (7.61)$$

119

The integral in [] is a one-dimensional integral because η is constant, so we can evaluate it using the one-dimensional Gauss quadrature analysis which we developed in the previous section. We use three integration points, labelled 1, 2 and 3 in Fig. 7.21. We shall see later that this choice leads to 9 integration points for the whole element.

We recall from eqn (7.55) of section 7.2.7 that the Gauss quadrature relation for integration using three integration points is

$$\int_{-1}^{+1} f(\xi, \eta).d\xi = w_1 f(\xi_1, \eta) + w_2 f(\xi_2, \eta) + w_3 f(\xi_3, \eta) \tag{7.62}$$

where $f(\xi_1, \eta)$ is the value of the function at Gauss point 1, $f(\xi_2, \eta)$ is the value of the function at Gauss point 2, and $f(\xi_3, \eta)$ is the value of the function at Gauss point 3.

The weights are $w_1 = 5/9$ $w_2 = 8/9$ and $w_3 = 5/9$.

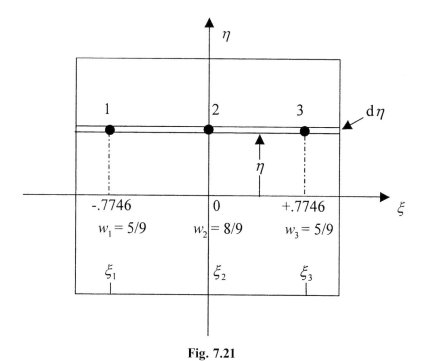

Fig. 7.21

The contribution to the integral from all the calculus elements in the strip of height $d\eta$ and length 2 in Fig. 7.21 is from eqn (7.62)

$$[w_1 f(\xi_1.\eta) + w_2 f(\xi_2, \eta) + w_3 f(\xi_3, \eta)].d\eta \tag{7.63}$$

To evaluate the complete integral, we need to sum eqn (7.63) over all the strips of area $2.d\eta$ from $\eta = -1$ to $\eta = +1$, that is

$$\int_{-1}^{1} [w_1 f(\xi_1, \eta) + w_2 f(\xi_2, \eta) + w_3 f(\xi_3, \eta)].d\eta$$

This is the sum of three integrals

$$\int_{-1}^{1} w_1 f(\xi_1, \eta).d\eta + \int_{-1}^{1} w_2 f(\xi_2, \eta).d\eta + \int_{-1}^{1} w_3 f(\xi_3, \eta).d\eta \tag{7.64}$$

Look at the first integral

$$\int_{-1}^{1} w_1 f(\xi_1, \eta).d\eta \tag{7.65}$$

This is a one-dimensional integral with η the single variable; w_1 and ξ_1 are constants. Let us be clear what this integral represents. The function $w_1 f(\xi_1, \eta).d\eta$ is the contribution to the overall integral from that part of the strip of total dimensions $2.d\eta$ associated with Gauss point 1 in Fig. 7.21 which extends from $\xi = -1$ to $\xi = -4/9$ – that is of length (weight) 5/9. So the integral (7.65) sums the contributions from all such strips from $\eta = -1$ to $\eta = +1$. This is described in the following diagram, which shows the pile of strips extending from $\eta = -1$ to $\eta = +1$ at $\xi = -0.7746$.

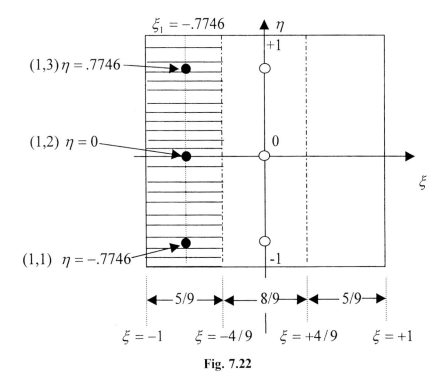

Fig. 7.22

We can sum the contribution from these strips by Gauss quadrature. The integral is along the line $\xi = \xi_1 = -0.7746$ from $\eta = -1$ to $+1$ shown in Fig. 7.22. We perform this integration using again three integration points which are labelled (1,1) (1,2) and (1,3) in Fig. 7.22. The first digit indicates that it is the first integration point for integrals along the ξ axis and the second digit is the order of the integration point for integrals along the η axis.

So the contribution to the total integral from the value of the integrated function over all the strips shown in Fig. 7.22 is (weight at Gauss point (1,1)). (value of the function at (1,1)) + (weight at Gauss point (1,2)). (value of the function at (1,2)) + (weight at Gauss point (1,3)). value of the function at (1,3)).

Recall that the function we are summing in this case is (see equation (7.65))

$$w_1 f(\xi_1, \eta)$$

So the Gaussian sum is

$$w_1.w_1 f(\xi_1, \eta_1) + w_2.w_1 f(\xi_1, \eta_2) + w_3.w_1 f(\xi_1, \eta_3) \tag{7.66}$$

121

Note that $f(\xi_1, \eta_2)$, for example, is the value of the function at $\xi = \xi_1$ and $\eta = \eta_2$, that is the point (1,2) in Fig. 7.22.

The second integral in eqn (7.64) is

$$\int_{-1}^{1} w_2 f(\xi_2.\eta).\mathrm{d}\eta$$

This is the contribution to the total integral from the parts of the $2.\mathrm{d}\eta$ strip in Fig. 7.21 associated with Gauss point 2, that is between $\xi = -4/9$ and $\xi = +4/9$. We can sum the contribution from all these strips by the same method that we used for the previous integral. Here we perform a Gauss quadrature sum along the line $\xi = \xi_2 = 0$. Again we use three integration points which are identified by open circles in Fig. 7.22. These integration points are also shown in Fig. 7.23, labelled, according to our convention, (2,1) (2,2) and (2,3) or b, e and h. Note that in this case the function we are integrating is $w_2 f(\xi_2, \eta)$. The Gauss quadrature value for the integral is

$$w_1.w_2 f(\xi_2, \eta_1) + w_2.w_2 f(\xi_2, \eta_2) + w_3.w_2 f(\xi_2.\eta_3) \tag{7.67}$$

Finally the third integral in eqn (7.64) is

$$\int_{-1}^{1} w_3 f(\xi_3, \eta).\mathrm{d}\eta$$

By arguments similar to those we have rehearsed, this represents the contribution to the total integral from the value of the function over the part of the strip in Fig 7.21 associated with the third integrating point, that is from $\xi = +4/9$ to $\xi = +1$. We sum the contribution from all such strips along the line $\xi = \xi_3$ from $\eta = -1$ to $\eta = +1$. The Gauss quadrature sum uses the integrating points (3,1) (3,2) and (3,3) or c, f and k in Fig. (7.23).The Gauss quadrature value is

$$w_1.w_3 f(\xi_3, \eta_1) + w_2 w_3 f(\xi_3, \eta_2) + w_3.w_3\, f(\xi_3, \eta_3) \tag{7.68}$$

We are now in a position to write down the value of the complete integral of eqn (7.64). It is the sum of eqns (7.66) (7.67) and (7.68), that is

$$w_1^2.f(\xi_1, \eta_1) + w_2 w_1.f(\xi_1, \eta_2) + w_3 w_1.f(\xi_1.\eta_3)+$$

$$w_1 w_2.f(\xi_2, \eta_1) + w_2^2.f(\xi_2, \eta_2) + w_3 w_2.f(\xi_2.\eta_3)+$$

$$w_1 w_3.f(\xi_3, \eta_1) + w_2 w_3.f(\xi_3.\eta_2) + w_3^2.f(\xi_3, \eta_3) \tag{7.69}$$

Each term in this set is the contribution to the overall integral from one Gauss point. It comprises the product of the value of the function at the Gauss point and a weight which is the product of the two one-dimensional weights corresponding to integration with respect to ξ and to η. The nine terms correspond to the nine integration points shown in Fig 7.23. The relation between the terms in (7.69) and the Gauss points is displayed in the following array which should be superimposed on equation (7.69),

$$a + d + g+$$
$$b + e + h+$$
$$c + f + k$$

For example, the fourth term in eqn (7.69) is the contribution from Gauss point b or (2,1) which has co-ordinates $\xi = \xi_2$ and $\eta = \eta_1$. The weight is the product of w_1 and w_2 and $f(\xi_2, \eta_1)$ is the value at Gauss point b of the function to be integrated. The weights of all the nine Gauss points are gathered together in Fig. 7.24.

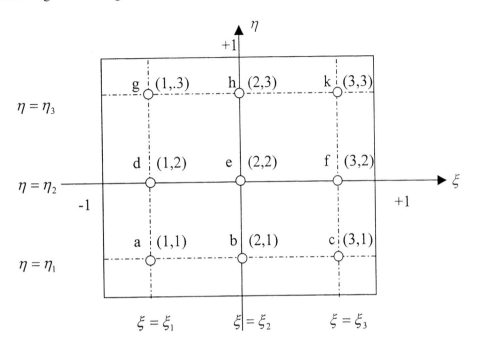

Fig. 7.23

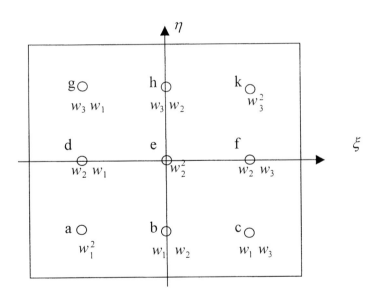

Fig. 7.24

We know that the one-dimensional weights are

$$w_1 = 5/9 \quad w_2 = 8/9 \quad \text{and} \quad w_3 = 5/9,$$

so the two-dimensional weights obtained by substituting these values are

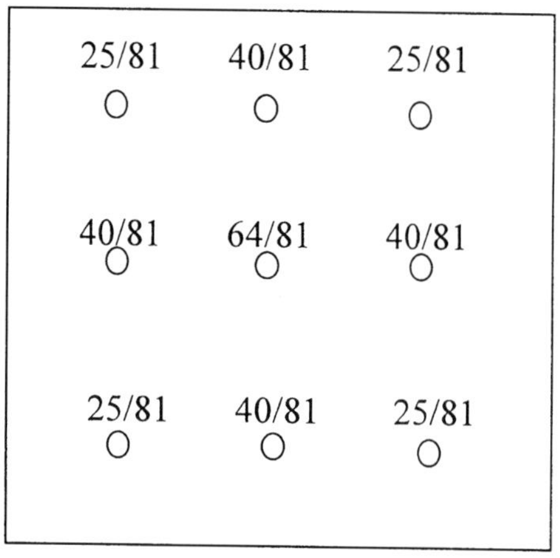

Fig. 7.25

We shall use these weights to perform the integration in one of the examples that follow.

The result that we have with much labour derived suggests a generalised procedure for obtaining the weights attached to a two-dimensional array of Gauss points. The two-dimensional array is set up by replicating orthogonally the Gauss points of the appropriate one-dimensional array. For example, the nine point array we have just analysed is derived from the three point one-dimensional set:

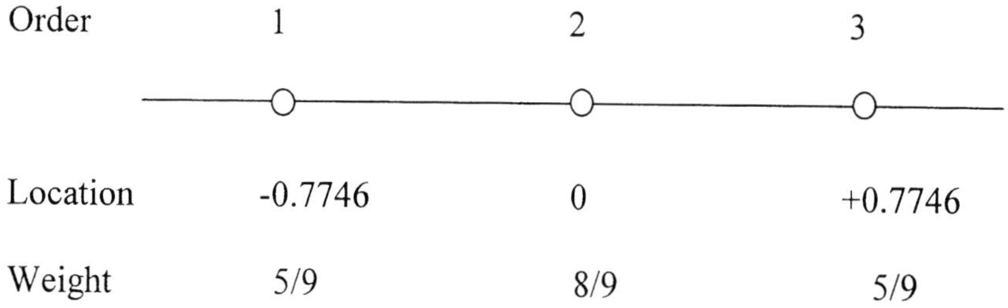

Order	1	2	3
	───O───	───O───	───O───
Location	-0.7746	0	+0.7746
Weight	5/9	8/9	5/9

It is evident from Fig. 7.24 that the combination of these weights for the Gauss points in the nine point array is given by the following scheme:

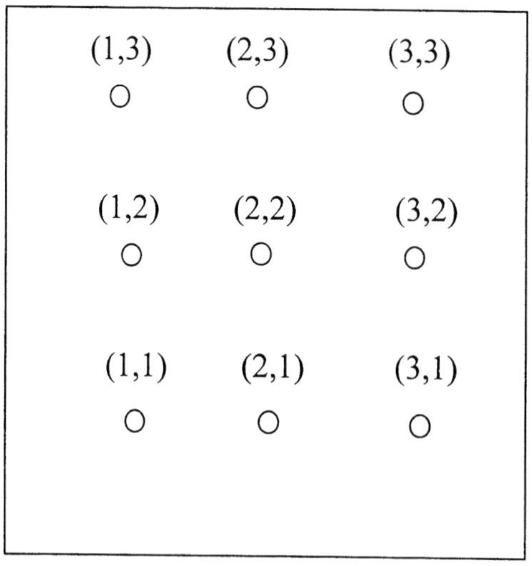

Fig. 7.26

124

By way of illustration we now apply this procedure to a 16-point array, which is based on the following four point one-dimensional set:

Order	1	2	3	4
Location	-0.8611	-0.3399	+0.3399	+0.8611
Weight	0.3479	0.6521	0.6521	0.3479

Note that the sum of the one-dimensional weights must always be 2.0, which is the length of the integration baseline. The pairing of the one-dimensional weights in the two-dimensional scheme follows the general pattern of the nine point array, that is

(1,4)	(2,4)	(3,4)	(4,4)
(1,3)	(2,3)	(3,3)	(4,3)
(1,2)	(2,2)	(3,2)	(4,2)
(1,1)	(2,1)	(3,1)	(4,1)

If we substitute the values for the one-dimensional weights in this scheme we obtain the following two-dimensional weights for the 16 point array

.1210	.2269	.2269	.1210
.2269	.4253	.4253	.2269
.2269	.4253	.4253	.2269
.1210	.2269	.2269	.1210

For example, the weight for point (1,1) is $(0.3479 \times 0.3479) = 0.1210$ the weight for point (2,4) is $(0.6521 \times 0.3479) = 0.2269$ and the weight for point (3,3) is $(0.6521 \times 0.6521) = 0.4253$.

Note that the symmetry of the Gauss point locations means that there are only three 'crystallographically distinct' points and therefore only three different weights. In general the points closest to the centre of the element have the highest weight.

Finally, we apply the principle to determine the weights of the very simple four point two-dimensional array which is derived from the two point one-dimensional set:

Order	1	2
Location	-0.5774	+0.5774
Weight	1.0	1.0

The pairing of these weights in the two-dimensional array is:

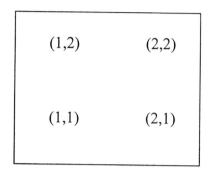

(1,2)	(2,2)
(1,1)	(2,1)

Since both one-dimensional weights are unity all the two-dimensional products are also unity, so the four point weights are:

1.0	1.0
1.0	1.0

We shall use this result in the next example. It is unlikely that the 16 point array would be used in finite element analysis.

APPENDIX 7.1   JUSTIFICATION AND PROOF THAT THE AREA OF A CALCULUS ELEMENT IN REAL $X-Y$ SPACE IS det J.d ξ.d η

We will first justify the validity of this relation as an interesting exercise in mapping, which will serve further to enhance our understanding of that process. This is followed by a more formal proof.

To the first end, we identify a small part P P_1 P_3 P_2 of the 8-node square master element in Fig. 7.1.1.

Locate the point P at $\xi = 0.25$, $\eta = 0.5$. Now translate P to P_1 a distance 0.01, which is dξ, so that ξ becomes 0.26. Then translate P to P_2 a distance 0.01, which is dη, so η becomes 0.51. Then PP_1 and PP_2 are two vectors that form the adjacent sides of a small square element. Complete the other two sides of the square by locating the point P_3 so PP_3 is the sum of the two vectors'. The co-ordinates of the corners of the square are listed in the table adjacent to the diagram.

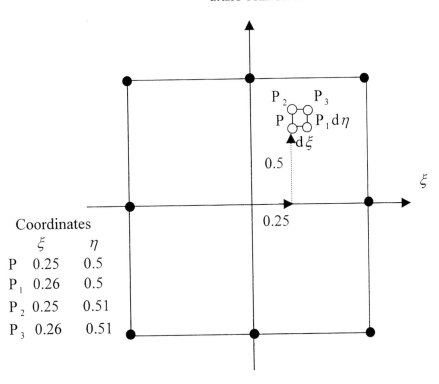

Fig. 7.1.1   Diagram of the Master Element identifying the points P P$_1$ P$_3$ P$_2$

We now map the displacements which form the sides of the small element on to the real element in x–y space. This will form the mapped dimensions of the small element in real space and we can then work out its area. We then compare this with the area of the corresponding element in the master element, which is evidently $(0.01 \times 0.01) = 0.0001$.

The displacement PP$_1$ in the master element is in the direction of ξ and is $d\xi = 0.01$. The corresponding displacement in the real element will have components in both the x direction (dx) and in the y direction (dy).

$$dx = \frac{\partial x}{\partial \xi} \cdot d\xi \text{ (i.e. the rate of change of } x \text{ with } \xi \text{ times the change of } \xi) \text{ and}$$

$$dy = \frac{\partial y}{\partial \xi} \cdot d\xi \text{ (i.e. the rate of change of } y \text{ with } \xi \text{ times the change of } \xi).$$

Now we know from the iso-parametric relation that

$$x = N_1 x_1 + N_2 x_2 + N_3 x_3 + N_4 x_4 + N_5 x_5 + N_6 x_6 + N_7 x_7 + N_8 x_8 \qquad (7.1.1)$$

where x_1 to x_8 are the nodal x-coordinates and N_1 to N_8 are the 8-node shape functions which are functions of ξ and η.
Differentiating (7.1.1) partially with respect to x

$$\frac{\partial x}{\partial \xi} = \frac{\partial N_1}{\partial \xi} x_1 + \frac{\partial N_2}{\partial \xi} x_2 + \frac{\partial N_3}{\partial \xi} x_3 + \frac{\partial N_4}{\partial \xi} x_4 + \frac{\partial N_5}{\partial \xi} x_5 + \frac{\partial N_6}{\partial \xi} x_6 + \frac{\partial N_7}{\partial \xi} x_7 + \frac{\partial N_8}{\partial \xi} x_8 \qquad (7.1.2)$$

because x_1 to x_8 are constants.

127

Similarly

$$y = N_1y_1 + N_2y_2 + N_3y_3 + N_4y_4 + N_5y_5 + N_6y_6 + N_7y_7 + N_8y_8 \qquad (7.1.3)$$

and

$$\frac{\partial y}{\partial \xi} = \frac{\partial N_1}{\partial \xi}y_1 + \frac{\partial N_2}{\partial \xi}y_2 + - - - - - - - - - - + \frac{\partial N_8}{\partial \xi}y_8 \qquad (7.1.4)$$

We will use eqns (7.1.2) and (7.1.4) to calculate the derivatives. The real element is drawn in Fig. (7.1.2). The nodal coordinates are

Node	x	y
1	10	20
2	70	10
3	50	60
4	20	50
5	40	15
6	60	35
7	35	55
8	15	35

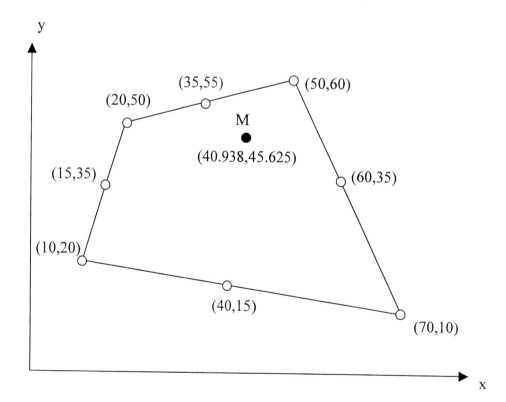

Fig. 7.1.2 Diagram of the Real Element showing the Nodal Coordinates and the location of M which is the mapped equivalent of P in the Master Element (Fig. 7.1.1)

128

We use the 8-node shape functions which are listed in Section (7.19) and the shape function derivatives which are listed below for ease of reference.

$$\frac{\partial N_1}{\partial \xi} = 0.25(1-\eta)(2\xi+\eta) \qquad \frac{\partial N_1}{\partial \eta} = 0.25(1-\xi)(2\eta+\xi)$$

$$\frac{\partial N_2}{\partial \xi} = 0.25(1-\eta)(2\xi-\eta) \qquad \frac{\partial N_2}{\partial \eta} = -0.25(1+\xi)(\xi-2\eta)$$

$$\frac{\partial N_3}{\partial \xi} = 0.25(1+\eta)(2\xi+\eta) \qquad \frac{\partial N_3}{\partial \eta} = 0.25(1+\eta)(2\eta+\xi)$$

$$\frac{\partial N_4}{\partial \xi} = 0.25(1+\eta)(2\xi-\eta) \qquad \frac{\partial N_4}{\partial \eta} = 0.25(1-\xi)(2\eta-\xi)$$

$$\frac{\partial N_5}{\partial \xi} = -\xi(1-\eta) \qquad \frac{\partial N_5}{\partial \eta} = -0.5(1-\xi^2)$$

$$\frac{\partial N_6}{\partial \xi} = 0.5(1-\eta^2) \qquad \frac{\partial N_6}{\partial \eta} = -\eta(1+\xi)$$

$$\frac{\partial N_7}{\partial \xi} = -\xi(1+\eta) \qquad \frac{\partial N_7}{\partial \eta} = 0.5(1-\xi^2)$$

$$\frac{\partial N_8}{\partial \xi} = -0.5(1-\eta^2) \qquad \frac{\partial N_8}{\partial \eta} = -\eta(1-\xi) \qquad (7.1.5)$$

First we use eqns (7.1.1) and (7.1.3) to establish the x and y coordinates of the point in the real element corresponding to the point P(0.25, 0.5) in the master element. Substituting these ξ and η coordinates in the shape function eqn (7.19) we get the following values

$$N_1 = -21/128 \qquad N_5 = 15/64$$
$$N_2 = -25/128 \qquad N_6 = 15/32$$
$$N_3 = -15/128 \qquad N_7 = 45/64$$
$$N_4 = -27/128 \qquad N_8 = 9/32$$

Note that these shape functions add up to unity – as they should. This is a useful check on the arithmetic.

So from eqns (7.1.1) and (7.1.3) the mapped x and y coordinates are

$$x = -(21/128 \times 10) - (25/128 \times 70) - (15/128 \times 50) - (27/128 \times 20) + (15/64 \times 40)$$
$$+ (15/32 \times 60) + (45/64 \times 35) + (9/32 \times 15) = \underline{40.938}$$
$$y = -(21/128 \times 20) - (25/128 \times 10) - (15/128 \times 60) - (27/128 \times 50) + (15/64 \times 15)$$
$$+ (15/32 \times 35) + (45/64 \times 55) + (9/32 \times 35) = \underline{45.625}$$

So the (x, y) coordinates corresponding to $\xi = 0.25$ and $\eta = 0.5$ are (40.938, 45.625). Call this point M. It is located in Fig. 7.1.2.

We now work out the movement of M in x–y space corresponding to the translation PP$_1$ in the master element. The translation in the x direction is

$$dx = \frac{\partial x}{\partial \xi}.d\xi$$

where $d\xi$ is the translation in the ξ direction in the master element. We get $\frac{\partial x}{\partial \xi}$ by substituting in eqn (7.1.2) the values of the shape function derivatives at $\xi = 0.25$ and $\eta = 0.5$ from eqn (7.1.5) and the real nodal coordinates. The terms in eqn (7.1.2) are listed in the following Table 7.1. Similarly, the translation in the y direction is

$$dy = \frac{\partial y}{\partial \xi}.d\xi$$

and is obtained similarly from eqn (7.1.4). The terms are again listed in the Table 7.1.

Table 7.1 Calculation of the displacement components in the real element corresponding to PP_1

Node i	Shape Function Derivative $\frac{\partial N_i}{\partial \xi}$	$x_i \frac{\partial N_i}{\partial \xi}$	$y_i \frac{\partial N_i}{\partial \xi}$
1	1/8	10/8	20/8
2	0	0	0
3	3/8	210/8	60/8
4	0	0	0
5	−1/8	−40/8	−15/8
6	3/8	180/8	35/8
7	−3/8	−105/8	−55/8
8	−3/8	−45/8	−35/8

$$\sum x_i \frac{\partial N_i}{\partial \xi} = \frac{\partial x}{\partial \xi} = 26.25 \qquad \sum y_i \frac{\partial N_i}{\partial \xi} = \frac{\partial y}{\partial \xi} = 1.25$$

So the x displacement is

$$dx = \frac{\partial x}{\partial \xi}.d\xi = 26.25 \times 0.01 = 0.2625 \text{ m}$$

and the y displacement is

$$dy = \frac{\partial y}{\partial \xi}d\xi = 1.25 \times 0.01 = 0.0125 \text{ m}$$

This gives the mapped point M_1 corresponding to P_1 and is identified in Fig. 7.1.3.

Similarly we can work out the translation in x–y space corresponding to the movement PP_2 in the master element, which is $d\eta = 0.01$. These mapped values of dx and dy in the real element are given in Table 7.2, which follows the same pattern as the previous one.

Table 7.2 Calculation of the displacement in the real element corresponding to PP_2

Node i	Shape Function Derivative $\frac{\partial N_i}{\partial \eta}$	$x_i \frac{\partial N_i}{\partial \eta}$	$y_i \frac{\partial N_i}{\partial \eta}$
1	15/64	150/64	300/64
2	15/64	1050/64	150/64
3	25/64	1250/64	1500/64
4	9/64	180/64	450/64
5	−15/32	−600/32	−225/32
6	−5/8	−300/8	−175/8
7	7/16	225/16	385/16
8	−3/8	−45/8	−105/8

$$\sum x_i \frac{\partial N_i}{\partial \eta} = \frac{\partial x}{\partial \eta} = -12.03 \qquad \sum y_i \frac{\partial N_i}{\partial \eta} = \frac{\partial y}{\partial \eta} = 19.53$$

So the x translation is $\mathrm{d}x = \dfrac{\partial x}{\partial \eta}\mathrm{d}\eta = -12.03 \times 0.01$

$$= -0.1203 \text{ m.}$$

and the y translation is $\mathrm{d}y = \dfrac{\partial y}{\partial \eta}\mathrm{d}\eta = 19.53 \times 0.01$

$$= 0.1953 \text{ m.}$$

This gives the position of M_2 corresponding to P_2 and is located in Fig. 7.1.3. We can now readily locate the position of M_3 corresponding to P_3 by completing the parallelogram in Fig. 7.1.3. This parallelogram is the mapped form of the square element $PP_1\, P_3\, P_2$ in the master element. We now work out its area.

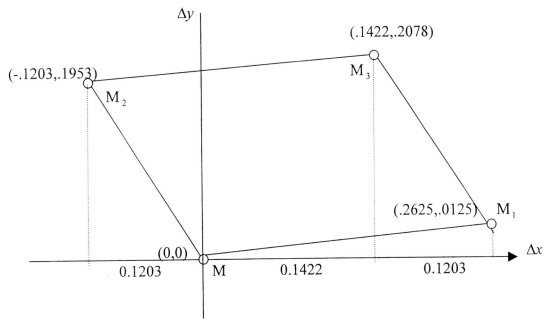

Fig. 7.1.3 The mapped form and dimensions in the real element of the element $PP_1P_3P_2$ in the master element (Fig. 7.1.2). The dimensions are defined in terms of distances $(\Delta x, \Delta y)$ from the point M. The co-ordinates of M in the real element are (40.938, 45.625)

Referring to the parallelogram-shaped element in Fig. 7.1.3, The area is the sum of two trapeziums minus the area of two triangles which is

$$(0.2625 - 0.1422)(0.5\{0.2078 + 0.0125\}) + (0.2625)(0.5\{0.2078 + 0.1953\})$$

$$- (0.5\{0.2625 \times 0.0125\}) - (0.5\{0.1203 \times 0.1953) = 0.05277 \text{ m}^2.$$

Now

$$\det J = \frac{\partial x}{\partial \xi}\frac{\partial y}{\partial \eta} - \frac{\partial x}{\partial \eta}\frac{\partial y}{\partial \xi}$$

$$= 26.25 \times 19.53 - (-12.03 \times 1.25)$$

$$= 527.7$$

and

$$\det J\mathrm{d}\,\xi.\mathrm{d}\eta = 527.7 \times 0.01 \times 0.01 = 0.05277 \text{ m}^2$$

which is equal to the area of the element in $x-y$ space, so the proposition is justified.

APPENDIX 7.2   A MORE FORMAL PROOF

Those privy to the mysteries of vector algebra would appreciate a more formal proof. Referring to Fig. 7.1.4, they would recognise the area of M M_1 M_3 M_2 as the vector product of the two vectors MM_1 and MM_2.

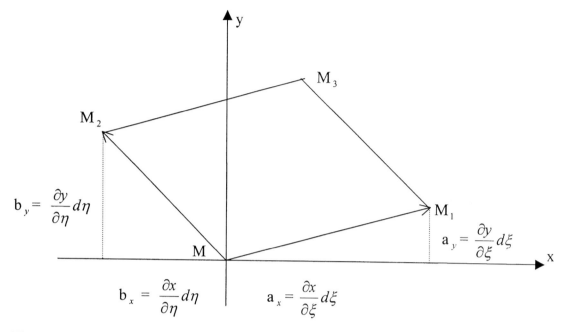

Fig. 7.1.4   Diagram showing the vectors MM_1 and MM_2 and the parallelogram MM_1 M_3 M_2 that is their vector product.

If we have two vectors

$$a_x i + a_y j \quad \text{and} \quad b_x i + b_y j$$

where i is a unit vector along the x-axis and j is a unit vector along the y-axis then the standard relation for the magnitude of the vector product of these two vectors is

$$a_x b_y - a_y b_x. \tag{7.1.6}$$

Referring again to Fig. 7.1.4 which represents our mapped real element,

$$a_x = \frac{\partial x}{\partial \xi} d\xi \quad a_y = \frac{\partial y}{\partial \xi} d\xi \quad b_x = \frac{\partial x}{\partial \eta} d\eta \quad b_y = \frac{\partial y}{\partial \eta} d\eta$$

Substituting these values in eqn (7.1.6) for the vector product, which is also the area of the parallelogram MM_1 M_3 M_2, yields

$$\left(\frac{\partial x}{\partial \xi} \frac{\partial y}{\partial \eta} - \frac{\partial x}{\partial \eta} \frac{\partial y}{\partial \xi} \right) d\xi . d\eta$$

We recognise the expression in ( ) as det J so the area of the real element is

$$\det J d\xi . d\eta$$

and the proposition is proved.

CHAPTER 8   WORKED EXAMPLES APPLYING THE THEORY OF SECTION (7.2) TO CALCULATE THE STRESSES IN A LOADED TAPERED SHEET

8.1   THE ANALYSIS USING FOUR INTEGRATION POINTS

8.1.1   OUTLINE OF THE EXAMPLE

The following example illustrates the procedure for calculating the stress distribution in an element using parabolic shape functions. These, as we have seen, give a linear distribution of stress over the element. The example we have chosen is a tapered sheet fixed along one edge and loaded on the opposite edge t. The dimensions of the sheet and the details of the loads and the constraints are given in Fig. 8.1.

The plate forms a single iso-parametric element with 8 nodes numbered 1 to 8. We shall perform the integration that leads to the stiffness matrix using four Gauss points, labelled a to d. Later in section (8.2) we shall analyse the same problem using nine Gauss points. We shall follow through the analysis step by step revealing the detailed structure of the analysis. However we shall not lay out the details of the matrix multiplication and matrix inversions used in the analysis. These are evaluated using the MATHCAD package. The calculations are worked through using 15 significant figures but the displayed numerical precision has been adjusted to fit the matrices into the page width.

This analysis based on a single element does not constitute a realistic engineering solution, which would use a mesh of a number of elements. However the critical part of the multi-element analysis is the derivation of the stiffness matrix of the individual element. This process is similar for all elements of a particular type. The combination of this set of similar stiffness matrices to form the global stiffness matrix of the stressed body involves the assembly of the stiffness matrices of the individual elements. The process of assembly for elements of the type we now analyse is identical in principle to that we used in the example in section (6.2.3), so developing the global stiffness matrix for a multi-element version of this problem presents no new principle.

By selecting a single element we reduce the arithmetical complexity and, in fact, produce a stress distribution which has some of the essential elements of the actual stress distribution.

8.1.2   STRATEGY FOR DETERMINING THE STIFFNESS MATRIX

The first phase of the analysis is to determine the nodal displacements. To this end we need the stiffness matrix of the element because this relates the applied nodal forces (F), which are known, to the nodal displacements (U) through the basic matrix equation

$$F = KU$$

We saw in Section 7.4 that the stiffness matrix is the result of integrating the matrix product

$$B^T \cdot D \cdot B \cdot \det J \cdot t$$

with respect to ξ and η over the area of the serendipity master element.

Finally we concluded that the above integral had to be evaluated numerically. We choose Gaussian quadrature with four gauss points. To this end we evaluate the matrix product

$$B^T \cdot D \cdot B \cdot \det J \cdot t$$

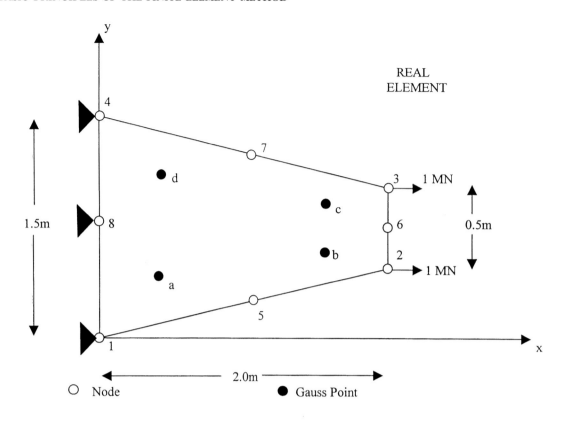

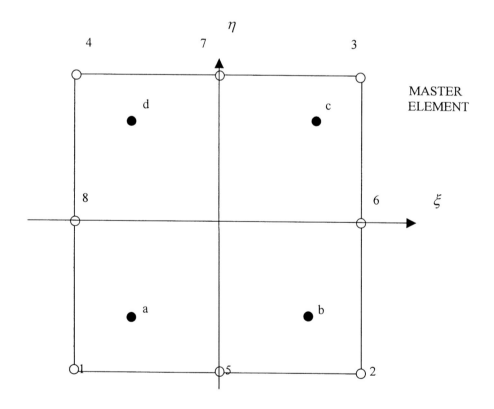

Fig. 8.1   The real element with four Gauss points and its master element

at the four Gauss points, multiply each product by the appropriate Gauss point weight and sum these products. This gives the stiffness matrix. We now proceed to perform this analysis.

8.1.3 Essential Data and Relationships needed for the evaluation of the Strain Matrix B

For this exercise we need the following data:

(i) The nodal co-ordinates in the real element
These are (in metres), where e.g. $x2$ is the x-coordinate of node 2, $y7$ is the y-coordinate of node 7 and so on. They are listed below.

Nodal Coordinates of the Real Element

$x1 := 0$	$y1 := 0$
$x2 := 2$	$y2 := 0.5$
$x3 := 2$	$y3 := 1$
$x4 := 0$	$y4 := 1.5$
$x5 := 1$	$y5 := 0.25$
$x6 := 2$	$y6 := 0.75$
$x7 := 1$	$y7 := 1.25$
$x8 := 0$	$y8 := 0.75$

The master element coordinates of the Gauss points are

Gauss Point	Xi	Eta
a	$-\dfrac{1}{\sqrt{3}}$	$-\dfrac{1}{\sqrt{3}}$
b	$\dfrac{1}{\sqrt{3}}$	$-\dfrac{1}{\sqrt{3}}$
c	$\dfrac{1}{\sqrt{3}}$	$\dfrac{1}{\sqrt{3}}$
d	$-\dfrac{1}{\sqrt{3}}$	$\dfrac{1}{\sqrt{3}}$

(ii) The matrix B is the product of the two matrices A and G
The details, relevant to this example, of the matrices A and G and of the product matrix B are given in section (7.2.4).

G contains the derivatives of the 8-node shape functions in the master element at the appropriate Gauss point. These derivatives are listed below, for ease of reference and for MATHCAD calculation purposes, in terms of the master element co-ordinates ξ and η. The 8-node shape functions and their derivatives are listed in eqns (7.19).

The derivatives are given the following symbols for computational purposes:

$s1$ is the value of the derivative of the node 1 shape function with respect to ξ, $s2$ is the derivative of the node 2 shape function and so on.
$t1$ is the derivative of the node 1 shape function with respect to η, $t2$ is the derivative of the node 2 shape function and so on.

135

At the head of the table, space is provided to insert the co-ordinates of a Gauss point. The MATHCAD programme then automatically calculates the values of the derivatives and of other related quantities at that Gauss point. The particular values displayed below are those for Gauss point *a*.

$$\xi := -\frac{1}{\sqrt{3}} \qquad\qquad \eta := -\frac{1}{\sqrt{3}}$$

Xi derivatives

$s1 := 0.25 \cdot (1 - \eta) \cdot (2 \cdot \xi + \eta)$
$s1 = -0.683013$
$s2 := 0.25 \cdot (1 - \eta) \cdot (2 \cdot \xi - \eta)$
$s2 = -0.227671$
$s3 := 0.25 \cdot (1 + \eta) \cdot (2 \cdot \xi + \eta)$
$s3 = -0.183013$
$s4 := 0.25 \cdot (1 + \eta) \cdot (2 \cdot \xi - \eta)$
$s4 = -0.061004$
$s5 := -\xi \cdot (1 - \eta)$
$s5 = 0.910684$
$s6 := 0.5 \cdot (1 - \eta^2)$
$s6 = 0.333333$
$s7 := -\xi \cdot (1 + \eta)$
$s7 = 0.244017$
$s8 := -0.5 \cdot (1 - \eta^2)$
$s8 = -0.333333$

Eta derivatives

$t1 := 0.25 \cdot (1 - \xi) \cdot (2 \cdot \eta + \xi)$
$t1 = -0.683013$
$t2 := -0.25 \cdot (1 + \xi) \cdot (\xi - 2 \cdot \eta)$
$t2 = -0.061004$
$t3 := 0.25 \cdot (1 + \xi) \cdot (2 \cdot \eta + \xi)$
$t3 = -0.183013$
$t4 := 0.25 \cdot (1 - \xi) \cdot (2 \cdot \eta - \xi)$
$t4 = -0.227671$
$t5 := -0.5 \cdot (1 - \xi^2)$
$t5 = -0.333333$
$t6 := -\eta \cdot (1 + \xi)$
$t6 = 0.244017$
$t7 := 0.5 \cdot (1 - \xi^2)$
$t7 = 0.333333$
$t8 := -\eta \cdot (1 - \xi)$
$t8 = 0.910684$

The *G* matrix, in terms of the above derivatives, is

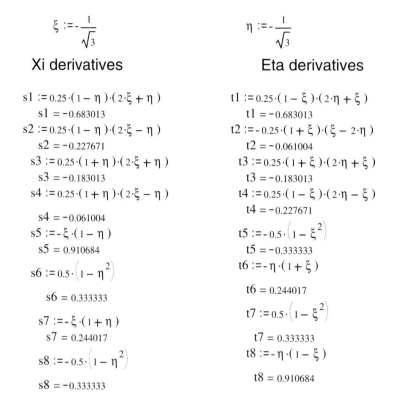

$$G := \begin{bmatrix} s1 & 0 & s2 & 0 & s3 & 0 & s4 & 0 & s5 & 0 & s6 & 0 & s7 & 0 & s8 & 0 \\ t1 & 0 & t2 & 0 & t3 & 0 & t4 & 0 & t5 & 0 & t6 & 0 & t7 & 0 & t8 & 0 \\ 0 & s1 & 0 & s2 & 0 & s3 & 0 & s4 & 0 & s5 & 0 & s6 & 0 & s7 & 0 & s8 \\ 0 & t1 & 0 & t2 & 0 & t3 & 0 & t4 & 0 & t5 & 0 & t6 & 0 & t7 & 0 & t8 \end{bmatrix}$$

The *G* matrix for Gauss point *a* is given the symbol *ga*, for Gauss point *b* it is *gb* and so on.

The matrix *A* contains the Jacobians J_{11}, J_{12}, J_{21} and J_{22} which are functions of the shape function derivatives and the nodal co-ordinates. The four Jacobians are defined below in terms of the computational symbols. These expressions then yield the values of the Jacobians when the nodal co-ordinates of the real element and the appropriate Gauss point master element co-ordinates are substituted. The numerical values printed below are for Gauss point *a*.

$$J_{11} := s1 \cdot x1 + s2 \cdot x2 + s3 \cdot x3 + s4 \cdot x4 + s5 \cdot x5 + s6 \cdot x6 + s7 \cdot x7 + s8 \cdot x8$$

$$J_{11} = 1.000000$$

$$J_{12} := s1 \cdot y1 + s2 \cdot y2 + s3 \cdot y3 + s4 \cdot y4 + s5 \cdot y5 + s6 \cdot y6 + s7 \cdot y7 + s8 \cdot y8$$

$$J_{12} = 0.435603$$

$$J_{21} := t1 \cdot x1 + t2 \cdot x2 + t3 \cdot x3 + t4 \cdot x4 + t5 \cdot x5 + t6 \cdot x6 + t7 \cdot x7 + t8 \cdot x8$$

$$J_{21} = 0.000000$$

$$J_{22} := t1 \cdot y1 + t2 \cdot y2 + t3 \cdot y3 + t4 \cdot y4 + t5 \cdot y5 + t6 \cdot y6 + t7 \cdot y7 + t8 \cdot y8$$

$$J_{22} = 0.644338$$

We also need the determinant of the Jacobian matrix. This is:

$$detJ := J_{11} \cdot J_{22} - J_{21} \cdot J_{12}$$

$$detJ = 0.644338$$

Let us pause for a moment to look at the value of $\det J$ that we have just calculated.

We recall from Appendix 7.1 that $\det J$ is the ratio of the area of a mapped calculus element in real space to the area of the calculus element in the master element from which it was mapped.

Now in this example the two elements have the same horizontal dimensions ($\xi = 2$ and $x = 2$) therefore a mapped calculus element will have the same horizontal dimension in the real element as that for the calculus element in the master element from which it was derived. Because of this fact, the ratio of the areas of the two calculus elements will be the ratio of the total heights of the two elements at the point in questions, which are not equal.

Looking at the particular case of Gauss point a, in the real elements its x-coordinate is $(1 - 1/\sqrt{3})$. The height of the element at this point is 1.288675 m. The height of the master element is 2 m. so the height ratio is

$$0.644338.$$

This is exactly the same as the calculated value of $\det J$ displayed above.

This simple test can be used to check values of $\det J$. As a further example, at node 1 the height of the real element is 1.5 m and of the master element is 2 m so we would expect that $\det J$ will be $1.5/2 = 0.75$ and $1/\det J = 1.33333$. We shall see later at the beginning of section (8.1.10) where we calculate the stresses at node 1 that it is. The calculated values of the Jacobians are substituted in the A matrix, which is

$$A := \frac{1}{detJ} \cdot \begin{bmatrix} J_{22} & -J_{12} & 0 & 0 \\ 0 & 0 & -J_{21} & J_{11} \\ -J_{21} & J_{11} & J_{22} & -J_{12} \end{bmatrix}$$

The last data we need are the elastic constants of the plate material which we choose to be those for those for aluminium. The following elasticity matrix D takes account of the fact that we

137

assume that the plate is in plane stress. The stress units are MPa.

$$D := 76150 \cdot \begin{bmatrix} 1 & .3 & 0 \\ .3 & 1 & 0 \\ 0 & 0 & .35 \end{bmatrix}$$

The plate thickness is

$$t := 0.02 \quad \text{metres}$$

8.1.4 CALCULATION OF THE MATRIX PRODUCT AT THE FOUR GAUSS POINTS

We are now in a position to calculate the matrix product

$$B^T \cdot D \cdot B \cdot \det J \cdot t$$

at the four Gauss points a, b, c and d in turn, which we now proceed to do. Note that B^T is the transpose of B, that is the B matrix with the rows and columns interchanged.

The matrix Product at Gauss Point a
The illustrative numbers displayed above are those for Gauss point a, so it is instructive to check that they appear correctly in the Gauss point a matrices that follows.

The A matrix is:

$$aa := 1.55198 \cdot \begin{bmatrix} .644338 & -.144338 & 0 & 0 \\ 0 & 0 & 0 & 1 \\ 0 & 1 & .644338 & -.144338 \end{bmatrix}$$

The g matrix is

$$ga := \begin{bmatrix} -.683 & 0 & -.22767 & 0 & -.183 & 0 & -.0610 & 0 & .91068 & 0 & .33333 & 0 & .2440 & 0 & -.33333 & 0 \\ -.683 & 0 & -.0610 & 0 & -.183 & 0 & -.22767 & 0 & -.33333 & 0 & .2440 & 0 & .33333 & 0 & .91068 & 0 \\ 0 & -.683 & 0 & -.22767 & 0 & -.183 & 0 & -.0610 & 0 & .91068 & 0 & .33333 & 0 & .2440 & 0 & -.33333 \\ 0 & -.683 & 0 & -.0610 & 0 & -.183 & 0 & -.2276 & 0 & -.33333 & 0 & .2440 & 0 & .33333 & 0 & .91068 \end{bmatrix}$$

The B matrix is the product of $aa \cdot ga$ and we designate this Ba

$$aa \cdot ga = \begin{bmatrix} -0.530 & 0.000 & -0.214 & 0.000 & -0.142 & 0.000 & -0.010 & 0.000 & 0.985 & 0.000 & 0.279 & 0.000 & 0.169 & 0.000 & -0.537 & 0.000 \\ 0.000 & -1.060 & 0.000 & -0.095 & 0.000 & -0.284 & 0.000 & -0.353 & 0.000 & -0.517 & 0.000 & 0.379 & 0.000 & 0.517 & 0.000 & 1.413 \\ -1.060 & -0.530 & -0.095 & -0.214 & -0.284 & -0.142 & -0.353 & -0.010 & -0.517 & 0.985 & 0.379 & 0.279 & 0.517 & 0.169 & 1.413 & -0.537 \end{bmatrix}$$

Note that this is a 3×16 matrix

$$Ba := aa \cdot ga$$

$$Ka := Ba^T \cdot D \cdot Ba \cdot .644338 \cdot .02$$

So Ka is the value of the matrix product at Gauss point a. Note that this is the product of three matrices of the following order – 16×3, 3×3 and 3×16. This yields a 16×16 matrix,

138

evaluated by MATHCAD below

$$
Ka = \begin{bmatrix}
661.6 & 358.4 & 145.8 & 92.7 & 177.3 & 96.0 & 133.8 & 58.8 & -324.1 & -278.0 & -282.8 & -160.5 & -276.4 & -142.4 & -235.1 & -24.9 \\
358.4 & 1199.1 & 84.0 & 137.4 & 96.0 & 321.3 & 67.4 & 369.3 & -213.3 & 358.8 & -155.9 & -444.6 & -147.0 & -568.9 & -89.6 & -1372.4 \\
145.8 & 84.0 & 48.0 & 12.9 & 39.1 & 22.5 & 13.6 & 22.6 & -190.1 & 0.6 & -70.8 & -32.9 & -52.4 & -38.1 & 66.9 & -71.6 \\
92.7 & 137.4 & 12.9 & 24.5 & 24.8 & 36.8 & 26.3 & 33.6 & 10.6 & -24.4 & -35.6 & -55.7 & -42.7 & -60.5 & -88.9 & -91.8 \\
177.3 & 96.0 & 39.1 & 24.8 & 47.5 & 25.7 & 35.9 & 15.7 & -86.8 & -74.5 & -75.8 & -43.0 & -74.1 & -38.1 & -63.0 & -6.7 \\
96.0 & 321.3 & 22.5 & 36.8 & 25.7 & 86.1 & 18.1 & 98.9 & -57.2 & 96.1 & -41.8 & -119.1 & -39.4 & -152.4 & -24.0 & -367.7 \\
133.8 & 67.4 & 13.6 & 26.3 & 35.9 & 18.1 & 43.0 & 2.3 & 53.1 & -118.1 & -48.7 & -34.9 & -64.4 & -22.1 & -166.3 & 61.0 \\
58.8 & 369.3 & 22.6 & 33.6 & 15.7 & 98.9 & 2.3 & 122.5 & -100.7 & 175.9 & -30.3 & -132.2 & -19.4 & -179.9 & 51.0 & -488.1 \\
-324.1 & -213.3 & -190.1 & 10.6 & -86.8 & -57.2 & 53.1 & -100.7 & 1044.7 & -325.1 & 202.2 & 60.3 & 71.8 & 120.0 & -770.7 & 505.5 \\
-278.0 & 358.8 & 0.6 & -24.4 & -74.5 & 96.1 & -118.1 & 175.9 & -325.1 & 596.1 & 85.7 & -97.9 & 149.3 & -205.3 & 560.2 & -899.4 \\
-282.8 & -155.9 & -70.8 & -35.6 & -75.8 & -41.8 & -48.7 & -30.3 & 202.2 & 85.7 & 125.5 & 67.3 & 113.6 & 64.5 & 36.9 & 46.1 \\
-160.5 & -444.6 & -32.9 & -55.7 & -43.0 & -119.1 & -34.9 & -132.2 & 60.3 & -97.9 & 67.3 & 167.4 & 68.4 & 208.5 & 75.4 & 473.8 \\
-276.4 & -147.0 & -52.4 & -42.7 & -74.1 & -39.4 & -64.4 & -19.4 & 71.8 & 149.3 & 113.6 & 68.4 & 120.1 & 55.9 & 161.8 & -25.0 \\
-142.4 & -568.9 & -38.1 & -60.5 & -38.1 & -152.4 & -22.1 & -179.9 & 120.0 & -205.3 & 64.5 & 208.5 & 55.9 & 272.5 & 0.4 & 686.3 \\
-235.1 & -89.6 & 66.9 & -88.9 & -63.0 & -24.0 & -166.3 & 51.0 & -770.7 & 560.2 & 36.9 & 75.4 & 161.8 & 0.4 & 969.4 & -484.4 \\
-24.9 & -1372.4 & -71.6 & -91.8 & -6.7 & -367.7 & 61.0 & -488.1 & 505.5 & -899.4 & 46.1 & 473.8 & -25.0 & 686.3 & -484.4 & 2059.4
\end{bmatrix}
$$

The Matrix Product for Gauss Point b

The *A* matrix is

$$
ab := 2.81166 \cdot \begin{bmatrix}
.355662 & -.144338 & 0 & 0 \\
0 & 0 & 0 & 1 \\
0 & 1 & .355662 & -.144338
\end{bmatrix}
$$

The *G* matrix is

$$
gb := \begin{bmatrix}
.22767 & 0 & .683 & 0 & .0610 & 0 & .183 & 0 & -.91068 & 0 & .33333 & 0 & -.244 & 0 & -.33333 & 0 \\
-.061 & 0 & -.683 & 0 & -.22767 & 0 & -.183 & 0 & -.33333 & 0 & .91068 & 0 & .33333 & 0 & .24401 & 0 \\
0 & .22767 & 0 & .683 & 0 & .0610 & 0 & .183 & 0 & -.91068 & 0 & .33333 & 0 & -.244 & 0 & -.33333 \\
0 & -.061 & 0 & -.683 & 0 & -.22767 & 0 & -.183 & 0 & -.33333 & 0 & .91068 & 0 & .33333 & 0 & .24401
\end{bmatrix}
$$

The *B* matrix is *Bb* and is $ab \cdot gb$

$$
ab \cdot gb = \begin{bmatrix}
0.252 & 0.000 & 0.960 & 0.000 & 0.153 & 0.000 & 0.257 & 0.000 & -0.775 & 0.000 & -0.036 & 0.000 & -0.379 & 0.000 & -0.432 & 0.000 \\
0.000 & -0.172 & 0.000 & -1.920 & 0.000 & -0.640 & 0.000 & -0.515 & 0.000 & -0.937 & 0.000 & 2.561 & 0.000 & 0.937 & 0.000 & 0.686 \\
-0.172 & 0.252 & -1.920 & 0.960 & -0.640 & 0.153 & -0.515 & 0.257 & -0.937 & -0.775 & 2.561 & -0.036 & 0.937 & -0.379 & 0.686 & -0.432
\end{bmatrix}
$$

$$Bb := ab \cdot gb$$

$$Kb := Bb^T \cdot D \cdot Bb \cdot .355662 \cdot .02$$

139

Kb is the value of the matrix product at Gauss point *b*.

$$Kb = \begin{bmatrix}
40.1 & -15.2 & 193.7 & -110.0 & 41.8 & -31.2 & 51.9 & -29.5 & -75.5 & -13.2 & -88.2 & 106.2 & -82.3 & 50.8 & -81.4 & 42.2 \\
-15.2 & 28.0 & -118.7 & 224.4 & -34.9 & 66.8 & -31.8 & 60.1 & -23.2 & 50.0 & 123.5 & -239.6 & 55.4 & -105.2 & 44.9 & -84.4 \\
193.7 & -118.7 & 1198.5 & -649.2 & 312.8 & -155.7 & 321.1 & -173.9 & -62.1 & 136.1 & -951.1 & 412.7 & -538.5 & 284.3 & -474.7 & 264.5 \\
-110.0 & 224.4 & -649.2 & 2172.4 & -164.4 & 693.8 & -173.9 & 582.1 & 71.4 & 833.7 & 477.4 & -2670.1 & 289.0 & -1043.9 & 259.8 & -792.4 \\
41.8 & -34.9 & 312.8 & -164.4 & 90.4 & -34.6 & 83.8 & -44.0 & 49.3 & 70.7 & -313.8 & 68.2 & -145.3 & 69.4 & -119.2 & 69.6 \\
-31.2 & 66.8 & -155.7 & 693.8 & -34.6 & 226.4 & -41.7 & 185.9 & 53.4 & 302.4 & 78.2 & -888.9 & 66.7 & -336.0 & 64.9 & -250.5 \\
51.9 & -31.8 & 321.1 & -173.9 & 83.8 & -41.7 & 86.0 & -46.6 & -16.6 & 36.5 & -254.8 & 110.6 & -144.3 & 76.2 & -127.2 & 70.9 \\
-29.5 & 60.1 & -173.9 & 582.1 & -44.0 & 185.9 & -46.6 & 156.0 & 19.1 & 223.4 & 127.9 & -715.4 & 77.4 & -279.7 & 69.6 & -212.3 \\
-75.5 & -23.2 & -62.1 & 71.4 & 49.3 & 53.4 & -16.6 & 19.1 & 492.2 & 255.9 & -439.7 & -316.2 & -7.2 & -50.7 & 59.7 & -9.6 \\
-13.2 & 50.0 & 136.1 & 833.7 & 70.7 & 302.4 & 36.5 & 223.4 & 255.9 & 589.8 & -370.9 & -1294.6 & -80.0 & -420.0 & -35.0 & -284.7 \\
-88.2 & 123.5 & -951.1 & 477.4 & -313.8 & 78.2 & -254.8 & 127.9 & -439.7 & -370.9 & 1243.7 & -32.7 & 462.4 & -189.6 & 341.5 & -213.9 \\
106.2 & -239.6 & 412.7 & -2670.1 & 68.2 & -888.9 & 110.6 & -715.4 & -316.2 & -1294.6 & -32.7 & 3551.6 & -164.3 & 1302.5 & -184.6 & 954.5 \\
-82.3 & 55.4 & -538.5 & 289.0 & -145.3 & 66.7 & -144.3 & 77.4 & -7.2 & -80.0 & 462.4 & -164.3 & 244.4 & -125.2 & 210.7 & -119.1 \\
50.8 & -105.2 & 284.3 & -1043.9 & 69.4 & -336.0 & 76.2 & -279.7 & -50.7 & -420.0 & -189.6 & 1302.5 & -125.2 & 503.1 & -115.2 & 379.4 \\
-81.4 & 44.9 & -474.7 & 259.8 & -119.2 & 64.9 & -127.2 & 69.6 & -35.0 & 341.5 & -184.6 & 210.7 & -115.2 & 190.5 & -104.4 \\
42.2 & -84.4 & 264.5 & -792.4 & 69.6 & -250.5 & 70.9 & -212.3 & -9.6 & -284.7 & -213.9 & 954.5 & -119.1 & 379.4 & -104.4 & 290.4
\end{bmatrix}$$

The Matrix Product at Gauss Point c

The *A* matrix is

$$ac := 2.81166 \cdot \begin{bmatrix}
.355662 & .144338 & 0 & 0 \\
0 & 0 & 0 & 1 \\
0 & 1 & .355662 & .144338
\end{bmatrix}$$

The *G* matrix is

$$gc := \begin{bmatrix}
.183 & 0 & .0610 & 0 & .683 & 0 & .22767 & 0 & -.2440 & 0 & .33333 & 0 & -.91068 & 0 & -.33333 & 0 \\
.183 & 0 & .22767 & 0 & .683 & 0 & .061 & 0 & -.33333 & 0 & -.91068 & 0 & .33333 & 0 & -.24401 & 0 \\
0 & .1830 & 0 & .0610 & 0 & .683 & 0 & .22767 & 0 & -.2440 & 0 & .33333 & 0 & -.91068 & 0 & -.33333 \\
0 & .1830 & 0 & .22767 & 0 & .683 & 0 & .061 & 0 & -.33333 & 0 & -.91068 & 0 & .33333 & 0 & -.24401
\end{bmatrix}$$

The *B* matrix is *Bc* and is *ac · gc*

$$ac \cdot gc = \begin{bmatrix}
0.257 & 0.000 & 0.153 & 0.000 & 0.960 & 0.000 & 0.252 & 0.000 & -0.379 & 0.000 & -0.036 & 0.000 & -0.775 & 0.000 & -0.432 & 0.000 \\
0.000 & 0.515 & 0.000 & 0.640 & 0.000 & 1.920 & 0.000 & 0.172 & 0.000 & -0.937 & 0.000 & -2.561 & 0.000 & 0.937 & 0.000 & -0.686 \\
0.515 & 0.257 & 0.640 & 0.153 & 1.920 & 0.960 & 0.172 & 0.252 & -0.937 & -0.379 & -2.561 & -0.036 & 0.937 & -0.775 & -0.686 & -0.432
\end{bmatrix}$$

$$Bc := ac \cdot gc$$

$$Kc := Bc^T \cdot D \cdot Bc \cdot .355662 \cdot .02$$

Kc is the value of the matrix product at Gauss point *c*

$$
Kc = \begin{bmatrix}
86.0 & 46.6 & 83.8 & 41.7 & 321.1 & 173.9 & 51.9 & 31.8 & -144.3 & -76.2 & -254.8 & -110.6 & -16.6 & -36.5 & -127.2 & -70.9 \\
46.6 & 156.0 & 44.0 & 185.9 & 173.9 & 582.1 & 29.5 & 60.1 & -77.4 & -279.7 & -127.9 & -715.4 & -19.1 & 223.4 & -69.6 & -212.3 \\
83.8 & 44.0 & 90.4 & 34.6 & 312.8 & 164.4 & 41.8 & 34.9 & -145.3 & -69.4 & -313.8 & -68.2 & 49.3 & -70.7 & -119.2 & -69.6 \\
41.7 & 185.9 & 34.6 & 226.4 & 155.7 & 693.8 & 31.2 & 66.8 & -66.7 & -336.0 & -78.2 & -888.9 & -53.4 & 302.4 & -64.9 & -250.5 \\
321.1 & 173.9 & 312.8 & 155.7 & 1198.5 & 649.2 & 193.7 & 118.7 & -538.5 & -284.3 & -951.1 & -412.7 & -62.1 & -136.1 & -474.7 & -264.5 \\
173.9 & 582.1 & 164.4 & 693.8 & 649.2 & 2172.4 & 110.0 & 224.4 & -289.0 & -1043.9 & -477.4 & -2670.1 & -71.4 & 833.7 & -259.8 & -792.4 \\
51.9 & 29.5 & 41.8 & 31.2 & 193.7 & 110.0 & 40.1 & 15.2 & -82.3 & -50.8 & -88.2 & -106.2 & -75.5 & 13.2 & -81.4 & -42.2 \\
31.8 & 60.1 & 34.9 & 66.8 & 118.7 & 224.4 & 15.2 & 28.0 & -55.4 & -105.2 & -123.5 & -239.6 & 23.2 & 50.0 & -44.9 & -84.4 \\
-144.3 & -77.4 & -145.3 & -66.7 & -538.5 & -289.0 & -82.3 & -55.4 & 244.4 & 125.2 & 462.4 & 164.3 & -7.2 & 80.0 & 210.7 & 119.1 \\
-76.2 & -279.7 & -69.4 & -336.0 & -284.3 & -1043.9 & -50.8 & -105.2 & 125.2 & 503.1 & 189.6 & 1302.5 & 50.7 & -420.0 & 115.2 & 379.4 \\
-254.8 & -127.9 & -313.8 & -78.2 & -951.1 & -477.4 & -88.2 & -123.5 & 462.4 & 189.6 & 1243.7 & 32.7 & -439.7 & 370.9 & 341.5 & 213.9 \\
-110.6 & -715.4 & -68.2 & -888.9 & -412.7 & -2670.1 & -106.2 & -239.6 & 164.3 & 1302.5 & 32.7 & 3551.6 & 316.2 & -1294.6 & 184.6 & 954.5 \\
-16.6 & -19.1 & 49.3 & -53.4 & -62.1 & -71.4 & -75.5 & 23.2 & -7.2 & 50.7 & -439.7 & 316.2 & 492.2 & -255.9 & 59.7 & 9.6 \\
-36.5 & 223.4 & -70.7 & 302.4 & -136.1 & 833.7 & 13.2 & 50.0 & 80.0 & -420.0 & 370.9 & -1294.6 & -255.9 & 589.8 & 35.0 & -284.7 \\
-127.2 & -69.6 & -119.2 & -64.9 & -474.7 & -259.8 & -81.4 & -44.9 & 210.7 & 115.2 & 341.5 & 184.6 & 59.7 & 35.0 & 190.5 & 104.4 \\
-70.9 & -212.3 & -69.6 & -250.5 & -264.5 & -792.4 & -42.2 & -84.4 & 119.1 & 379.4 & 213.9 & 954.5 & 9.6 & -284.7 & 104.4 & 290.4
\end{bmatrix}
$$

The Matrix Product at Gauss Point d

The *A* matrix is *ad*

$$
ad := 1.55198 \cdot \begin{bmatrix}
.644338 & .144337 & 0 & 0 \\
0 & 0 & 0 & 1 \\
0 & 1 & .644338 & .144337
\end{bmatrix}
$$

The *G* matrix is *gd*

$$
gd := \begin{bmatrix}
-.0610 & 0 & -.183 & 0 & -.22767 & 0 & -.683 & 0 & .2440 & 0 & .33333 & 0 & .91068 & 0 & -.33333 & 0 \\
.22767 & 0 & .183 & 0 & .06100 & 0 & .683 & 0 & -.33333 & 0 & -.2440 & 0 & .33333 & 0 & -.91068 & 0 \\
0 & -.0610 & 0 & -.183 & 0 & -.22767 & 0 & -.683 & 0 & .2440 & 0 & .33333 & 0 & .91068 & 0 & -.33333 \\
0 & .22767 & 0 & .183 & 0 & .06100 & 0 & .683 & 0 & -.33333 & 0 & -.2440 & 0 & .33333 & 0 & -.91068
\end{bmatrix}
$$

The *B* matrix is *Bd* and is *ad · gd*

$$
ad \cdot gd = \begin{bmatrix}
-0.010 & 0.000 & -0.142 & 0.000 & -0.214 & 0.000 & -0.530 & 0.000 & 0.169 & 0.000 & 0.279 & 0.000 & 0.985 & 0.000 & -0.537 & 0.000 \\
0.000 & 0.353 & 0.000 & 0.284 & 0.000 & 0.095 & 0.000 & 1.060 & 0.000 & -0.517 & 0.000 & -0.379 & 0.000 & 0.517 & 0.000 & -1.413 \\
0.353 & -0.010 & 0.284 & -0.142 & 0.095 & -0.214 & 1.060 & -0.530 & -0.517 & 0.169 & -0.379 & 0.279 & 0.517 & 0.985 & -1.413 & -0.537
\end{bmatrix}
$$

$$Bd := ad \cdot gd$$

$$Kd := Bd^{T} \cdot D \cdot Bd \cdot .644338 \cdot .02$$

Kd is the value of the matrix product at Gauss point d

$$Kd = \begin{bmatrix}
43.0 & -2.3 & 35.9 & -18.1 & 13.6 & -26.3 & 133.8 & -67.4 & -64.4 & 22.1 & -48.7 & 34.9 & 53.1 & 118.1 & -166.3 & -61.0 \\
-2.3 & 122.6 & -15.7 & 99.0 & -22.6 & 33.6 & -58.8 & 369.4 & 19.4 & -180.0 & 30.3 & -132.3 & 100.7 & 176.0 & -51.0 & -488.2 \\
35.9 & -15.7 & 47.5 & -25.7 & 39.1 & -24.8 & 177.3 & -96.0 & -74.1 & 38.1 & -75.8 & 43.0 & -86.8 & 74.5 & -63.0 & 6.7 \\
-18.1 & 99.0 & -25.7 & 86.1 & -22.5 & 36.8 & -96.0 & 321.3 & 39.4 & -152.4 & 41.8 & -119.1 & 57.2 & 96.1 & 24.0 & -367.7 \\
13.6 & -22.6 & 39.1 & -22.5 & 48.0 & -12.9 & 145.8 & -84.0 & -52.4 & 38.1 & -70.8 & 32.9 & -190.1 & -0.6 & 66.9 & 71.6 \\
-26.3 & 33.6 & -24.8 & 36.8 & -12.9 & 24.5 & -92.7 & 137.4 & 42.7 & -60.5 & 35.6 & -55.7 & -10.6 & -24.4 & 88.9 & -91.8 \\
133.8 & -58.8 & 177.3 & -96.0 & 145.8 & -92.7 & 661.6 & -358.4 & -276.4 & 142.4 & -282.8 & 160.5 & -324.1 & 278.0 & -235.1 & 24.9 \\
-67.4 & 369.4 & -96.0 & 321.3 & -84.0 & 137.4 & -358.4 & 1199.1 & 147.0 & -568.9 & 155.9 & -444.6 & 213.3 & 358.8 & 89.6 & -1372.4 \\
-64.4 & 19.4 & -74.1 & 39.4 & -52.4 & 42.7 & -276.4 & 147.0 & 120.1 & -55.9 & 113.6 & -68.4 & 71.8 & -149.3 & 161.8 & 25.0 \\
22.1 & -180.0 & 38.1 & -152.4 & 38.1 & -60.5 & 142.4 & -568.9 & -55.9 & 272.5 & -64.5 & 208.5 & -120.0 & -205.3 & -0.4 & 686.3 \\
-48.7 & 30.3 & -75.8 & 41.8 & -70.8 & 35.6 & -282.8 & 155.9 & 113.6 & -64.5 & 125.5 & -67.3 & 202.2 & -85.7 & 36.9 & -46.1 \\
34.9 & -132.3 & 43.0 & -119.1 & 32.9 & -55.7 & 160.5 & -444.6 & -68.4 & 208.5 & -67.3 & 167.4 & -60.3 & -97.9 & -75.4 & 473.8 \\
53.1 & 100.7 & -86.8 & 57.2 & -190.1 & -10.6 & -324.1 & 213.3 & 71.8 & -120.0 & 202.2 & -60.3 & 1044.7 & 325.1 & -770.7 & -505.5 \\
118.1 & 176.0 & 74.5 & 96.1 & -0.6 & -24.4 & 278.0 & 358.8 & -149.3 & -205.3 & -85.7 & -97.9 & 325.1 & 596.1 & -560.2 & -899.4 \\
-166.3 & -51.0 & -63.0 & 24.0 & 66.9 & 88.9 & -235.1 & 89.6 & 161.8 & -0.4 & 36.9 & -75.4 & -770.7 & -560.2 & 969.4 & 484.4 \\
-61.0 & -488.2 & 6.7 & -367.7 & 71.6 & -91.8 & 24.9 & -1372.4 & 25.0 & 686.3 & -46.1 & 473.8 & -505.5 & -899.4 & 484.4 & 2059.4
\end{bmatrix}$$

8.1.5 THE ELEMENT STIFFNESS MATRIX

We have now created the four matrix products at the four Gauss points. Note that they are all 16×16 matrices and are all symmetrical, that is terms in mirror image positions about the main diagonal are equal.

To obtain the stiffness matrix K for the whole element by Gauss quadrature, we sum the products of the matrix product and the weight at each point. We saw in the previous section that for four integration points the weights are all unity so K is simply the sum of the four matrix products. That is:

$$K := Ka + Kb + Kc + Kd$$

All the summed matrices are 16×16 and the matrix summation process involves adding the four terms in corresponding positions in the four matrices to yield the term in the same position in the final K matrix. So, for example, the term in the top left-hand corner of the K matrix is

$$661.6 + 40.1 + 86.0 + 43.0 = 830.7$$
$$Ka \qquad Kb \qquad Kc \qquad Kd \qquad K$$

The final stiffness matrix K for the element is printed below. It is displayed to one decimal place although it is calculated to more than this. The first term in the top left-hand corner is 830.7 which accords with the above sum.

$$
K = \begin{bmatrix}
830.7 & 387.5 & 459.2 & 6.3 & 553.8 & 212.5 & 371.5 & -6.4 & -608.4 & -345.4 & -674.5 & -130.0 & -322.3 & -10 & -609.9 & -114.6 \\
387.5 & 1505.6 & -6.3 & 646.7 & 212.5 & 1003.7 & 6.3 & 858.9 & -294.6 & -51.0 & -130.0 & -1531.9 & -10 & -274.8 & -165.4 & -2157.3 \\
459.2 & -6.3 & 1384.5 & -627.4 & 703.8 & 6.3 & 553.8 & -212.5 & -471.5 & 105.4 & -1411.4 & 354.6 & -628.4 & 250.0 & -589.9 & 130.0 \\
6.3 & 646.7 & -627.4 & 2509.4 & -6.3 & 1461.2 & -212.5 & 1003.7 & 54.6 & 320.9 & 405.4 & -3733.8 & 250.0 & -705.9 & 130.0 & -1502.3 \\
553.8 & 212.5 & 703.8 & -6.3 & 1384.5 & 627.4 & 459.2 & 6.3 & -628.4 & -250.0 & -1411.4 & -354.6 & -471.5 & -105.4 & -589.9 & -130.0 \\
212.5 & 1003.7 & 6.3 & 1461.2 & 627.4 & 2509.4 & -6.3 & 646.6 & -250.0 & -705.9 & -405.4 & -3733.8 & -54.6 & 320.9 & -130.0 & -1502.3 \\
371.5 & 6.3 & 553.8 & -212.5 & 459.2 & -6.3 & 830.7 & -387.5 & -322.3 & 10' & -674.5 & 130.0 & -608.4 & 345.4 & -610.0 & 114.6 \\
-6.4 & 858.9 & -212.5 & 1003.7 & 6.3 & 646.6 & -387.5 & 1505.5 & 10.0 & -274.8 & 130.0 & -1531.9 & 294.6 & -50.9 & 165.3 & -2157.2 \\
-608.4 & -294.6 & -471.5 & 54.6 & -628.4 & -250.0 & -322.3 & 10.0 & 1901.4 & 0.0 & 338.4 & -160.0 & 129.2 & 0.0 & -338.4 & 640.0 \\
-345.4 & -51.0 & 105.4 & 320.9 & -250.0 & -705.9 & 10 & -274.8 & 0.0 & 1961.4 & -160.0 & 118.5 & 0.0 & -1250.7 & 640.0 & -118.5 \\
-674.5 & -130.0 & -1411.4 & 405.4 & -1411.4 & -405.4 & -674.5 & 130.0 & 338.4 & -160.0 & 2738.3 & -9.1 \bullet 10^{-5} & 338.4 & 160.0 & 756.8 & -0.0 \\
-130.0 & -1531.9 & 354.6 & -3733.8 & -354.6 & -3733.8 & 130.0 & -1531.9 & -160.0 & 118.5 & -9.1 \bullet 10^{-5} & 7438.0 & 160.0 & 118.5 & -0.0 & 2856.6 \\
-322.3 & -10 & -628.4 & 250.0 & -471.5 & -54.6 & -608.4 & 294.6 & 129.2 & 0.0 & 338.4 & 160.0 & 1901.4 & -0.0 & -338.4 & -640.0 \\
-10 & -274.8 & 250.0 & -705.9 & -105.4 & 320.9 & 345.4 & -50.9 & 0.0 & -1250.7 & 160.0 & 118.5 & -0.0 & 1961.4 & -640.0 & -118.5 \\
-609.9 & -165.4 & -589.9 & 130.0 & -589.9 & -130.0 & -610.0 & 165.3 & -338.4 & 640.0 & 756.8 & -0.0 & -338.4 & -640.0 & 2319.8 & -0.0 \\
-114.6 & -2157.3 & 130.0 & -1502.3 & -130.0 & -1502.3 & 114.6 & -2157.2 & 640.0 & -118.5 & -0.0 & 2856.6 & -640.0 & -118.5 & -0.0 & 4699.7
\end{bmatrix}
$$

$$(8.1.1)$$

With the evaluation of the ELEMENT STIFFNESS MATRIX we have achieved our first objective. Note that this matrix is 16×16 and is symmetrical about the major diagonal.

It is helpful at this stage to remind ourselves of the structure of the stiffness matrix equation. It relates the 16 force components at the 8 nodes to the 16 corresponding nodal displacements. The structure of the equation is

	u_1	v_1	u_2	v_2	u_3	v_3	u_4	v_4	u_5	v_5	u_6	v_6	u_7	v_7	u_8	v_8	
F_{u1}	O	O	O	O	O	O	O	O	O	O	O	O	O	O	O	O	u_1
F_{v1}	O	O	O	O	O	O	O	O	O	O	O	O	O	O	O	O	v_1
F_{u2}	O	O	O	O	O	O	O	O	O	O	O	O	O	O	O	O	u_2
F_{v2}	O	O	O	O	O	O	O	O	O	O	O	O	O	O	O	O	v_2
F_{u3}	O	O	O	O	O	O	O	O	O	O	O	O	O	O	O	O	u_3
F_{v3}	O	O	O	O	O	O	O	O	O	O	O	O	O	O	O	O	v_3
F_{u4}	O	O	O	O	O	O	O	O	O	O	O	O	O	O	O	O	u_4
F_{v4} =	O	O	O	O	O	O	O	O	O	O	O	O	O	O	O	O	v_4
F_{u5}	O	O	O	O	O	O	O	O	O	O	O	O	O	O	O	O	u_5
F_{v5}	O	O	O	O	O	O	O	O	O	O	O	O	O	O	O	O	v_5
F_{u6}	O	O	O	O	O	O	O	O	O	O	O	O	O	O	O	O	u_6
F_{v6}	O	O	O	O	O	O	O	O	O	O	O	O	O	O	O	O	v_6
F_{u7}	O	O	O	O	O	O	O	O	O	O	O	O	O	O	O	O	u_7
F_{v7}	O	O	O	O	O	O	O	O	O	O	O	O	O	O	O	O	v_7
F_{u8}	O	O	O	O	O	O	O	O	O	O	O	O	O	O	O	O	u_8
F_{v8}	O	O	O	O	O	O	O	O	O	O	O	O	O	O	O	O	v_8
Nodal Force Vector					Stiffness Matrix								Nodal Displacement Vector				

$$(8.1.2)$$

143

The first column is the set of 16 force components called the Nodal Force Vector. The component F_{v4} for example is the vertical force at node 4. The last column is the set of 16 nodal displacements, called the Nodal Displacement Vector. For example u_7 is the horizontal displacement at node 7.

The 16×16 set of 256 coefficients, indicated by 'o's, is the Stiffness Matrix K. The numerical values for the coefficients in the present example are printed in the K matrix above.

We recall from Chapter 3 that (8.1.2) represents 16 simultaneous equations, one for each force component. For example, the equation for F_{u1} us extracted by multiplying the coefficients in row 1 by the appropriate nodal displacement, that is

the coefficient in column 1 multiplies $u1$
the coefficient in column 2 multiplies $v1$
the coefficient in column 3 multiplies $u2$
and so on.

Similarly the equation for F_{v1} is obtained by multiplying the coefficients in row 2 by the nodal displacements according to the same scheme.

To emphasise which stiffness coefficient multiplies each nodal displacement to extract the force equations, we have indicated the appropriate nodal displacement at the top of the columns in the stiffness matrix.

8.1.6   The Condensed Stiffness Matrix

Our objective is to obtain values for the 10 unknown nodal displacements at nodes 2, 3, 5, 6 and 7. The displacements at nodes 1, 4 and 8 are known. They are all zero because we have fixed these nodes. To obtain the 10 unknown displacements we need 10 equations in which the only unknowns are the 10 sought displacements. We know all the 10 applied forces at nodes 2, 3, 5, 6 and 7. They are all zero except F_{u2} and F_{u3}, which are both 1MN. We do not know the forces at nodes 1, 4 and 8. These are the forces needed to fix these nodes. We return to them later.

If we choose the following equations

for F_{u2}, F_{v2}, F_{u3}, F_{v3}, F_{u5}, F_{v5}, F_{u6}, F_{v6}, F_{u7} and F_{v7}

which are formed from rows 3, 4, 5, 6, 9, 10, 11, 12, 13 and 14 then we shall have the required equations with the sought 10 unknowns. We achieve this end by deleting the rows corresponding to the unwanted equations in the K matrix, that is rows 1, 2, 7, 8, 15 and 16.

When we examine these residual equations we realise that the stiffness coefficients that multiply the nodal displacements u_1, v_1, u_4, v_4, u_8 and v_8 yield zero terms because all these displacements are zero. These coefficients are all those that lie in column 1, 2, 7, 8, 15 and 16, so we lose nothing by deleting these coefficients and the displacements that they multiply.

The consequence of making these two sets of deletions of rows and columns is a matrix equation with 10 force components and 10 nodal displacements linked by a 10×10 stiffness matrix. This is called the condensed or reduced matrix. We give it the symbol R.

When we make these deletions from the K matrix of eqn (8.1.1) we get the following reduced matrix

$$
\begin{bmatrix}
1384.5 & -627.4 & 703.8 & 6.3 & -471.5 & 105.4 & -1411.4 & 354.6 & -628.4 & 250.0 \\
-627.4 & 2509.4 & -6.3 & 1461.2 & 54.6 & 320.9 & 405.4 & -3733.8 & 250.0 & -705.9 \\
703.8 & -6.3 & 1384.5 & 627.4 & -628.4 & -250.0 & -1411.4 & -354.6 & -471.5 & -105.4 \\
6.3 & 1461.2 & 627.4 & 2509.4 & -250.0 & -705.9 & -405.4 & -3733.8 & -54.6 & 320.9 \\
-471.5 & 54.6 & -628.4 & -250.0 & 1901.4 & 0.0 & 338.4 & -160.0 & 129.2 & 0.0 \\
105.4 & 320.9 & -250.0 & -705.9 & 0.0 & 1961.4 & -160.0 & 118.5 & 0.0 & -1250.7 \\
-1411.4 & 405.4 & -1411.4 & -405.4 & 338.4 & -160.0 & 2738.3 & -0.0 & 338.4 & 160.0 \\
354.6 & -3733.8 & -354.6 & -3733.8 & -160.0 & 118.5 & -0.0 & 7438.0 & 160.0 & 118.5 \\
-628.4 & 250.0 & -471.5 & -54.6 & 129.2 & 0.0 & 338.4 & 160.0 & 1901.4 & -0.0 \\
250.0 & -705.9 & -105.4 & 320.9 & 0.0 & -1250.7 & 160.0 & 118.5 & -0.0 & 1961.4
\end{bmatrix}
$$

The reduced matrix equation is therefore

$$
\begin{bmatrix}
F_{u2} \\ F_{v2} \\ F_{u3} \\ F_{v3} \\ F_{u5} \\ F_{v5} \\ F_{u6} \\ F_{v6} \\ F_{u7} \\ F_{v7}
\end{bmatrix}
=
\begin{bmatrix}
1384.5 & -627.4 & 703.7 & 6.4 & -471.4 & 105.4 & -1411.4 & 354.6 & -628.4 & 250.0 \\
-627.4 & 2509.4 & -6.3 & 1461.0 & 54.6 & 321.0 & 405.4 & -3733.8 & 250.0 & -705.9 \\
703.7 & -6.3 & 1384.4 & 627.5 & -628.4 & -250.0 & -1411.3 & -354.6 & -471.5 & -105.4 \\
6.4 & 1461.0 & 627.5 & 2509.3 & -250.0 & -706.0 & -405.4 & -3733.5 & -54.6 & 321.1 \\
-471.4 & 54.6 & -628.4 & -250.0 & 1901.2 & 0.1 & 338.4 & -160.0 & 129.2 & -0.0 \\
105.4 & 321.0 & -250.0 & -706.0 & 0.1 & 1961.3 & -160.0 & 118.5 & -0.0 & -1250.7 \\
-1411.4 & 405.4 & -1411.3 & -405.4 & 338.4 & -160.0 & 2738.3 & -0.0 & 338.4 & 160.0 \\
354.6 & -3733.8 & -354.6 & -3733.5 & -160.0 & 118.5 & -0.0 & 7438.0 & 160.0 & 118.4 \\
-628.4 & 250.0 & -471.5 & -54.6 & 129.2 & -0.0 & 338.4 & 160.0 & 1901.4 & -0.0 \\
250.0 & -705.9 & -105.4 & 321.1 & -0.0 & -1250.7 & 160.0 & 118.4 & -0.0 & 1961.4
\end{bmatrix}
\begin{bmatrix}
u_2 \\ v_2 \\ u_3 \\ v_3 \\ u_5 \\ v_5 \\ u_6 \\ v_6 \\ u_7 \\ v_7
\end{bmatrix}
$$

For computational purposes we have defined the condensed matrix by R so

$$
R :=
\begin{bmatrix}
1384.5 & -627.4 & 703.8 & 6.3 & -471.5 & 105.4 & -1411.4 & 354.6 & -628.4 & 250.0 \\
-627.4 & 2509.4 & -6.3 & 1461.2 & 54.6 & 320.9 & 405.4 & -3733.8 & 250.0 & -705.9 \\
703.8 & -6.3 & 1384.5 & 627.4 & -628.4 & -250.0 & -1411.4 & -354.6 & -471.5 & -105.4 \\
6.3 & 1461.2 & 627.4 & 2509.4 & -250.0 & -705.9 & -405.4 & -3733.8 & -54.6 & 320.9 \\
-471.5 & 54.6 & -628.4 & -250.0 & 1901.4 & 0.0 & 338.4 & -160.0 & 129.2 & 0.0 \\
105.4 & 320.9 & -250.0 & -705.9 & 0.0 & 1961.4 & -160.0 & 118.5 & 0.0 & -1250.7 \\
-1411.4 & 405.4 & -1411.4 & -405.4 & 338.4 & -160.0 & 2738.3 & -0.0 & 338.4 & 160.0 \\
354.6 & -3733.8 & -354.6 & -3733.8 & -160.0 & 118.5 & -0.0 & 7438.0 & 160.0 & 118.5 \\
-628.4 & 250.0 & -471.5 & -54.6 & 129.2 & 0.0 & 338.4 & 160.0 & 1901.4 & -0.0 \\
250.0 & -705.9 & -105.4 & 320.9 & 0.0 & -1250.7 & 160.0 & 118.5 & -0.0 & 1961.4
\end{bmatrix}
$$

145

8.1.7 INVERSION OF THE CONDENSED STIFFNESS MATRIX TO DETERMINE THE NODAL DISPLACEMENTS

The reduced stiffness matrix equation has the form

$$F = RU$$

Where F is the force vector and U is the displacement vector. To get the nodal displacements we need to invert this equation to

$$F = R^{-1}U$$

Inverting R using the MATHCAD package yields

$$R^{-1} = \begin{bmatrix}
0.003516 & 0.003956 & -0.000330 & 0.003734 & 0.001187 & 0.000616 & 0.001474 & 0.003686 & 1.424064 \bullet 10^{-5} & 0.000397 \\
0.003956 & 0.018452 & -0.003734 & 0.016612 & 0.003100 & 0.004405 & -0.000181 & 0.017220 & -0.003195 & 0.005001 \\
-0.000330 & -0.003734 & 0.003516 & -0.003956 & 1.424064 \bullet 10^{-5} & -0.000397 & 0.001474 & -0.003686 & 0.001187 & -0.000616 \\
0.003734 & 0.016612 & -0.003956 & 0.018452 & 0.003195 & 0.005001 & 0.000181 & 0.017220 & -0.003100 & 0.004405 \\
0.001187 & 0.003100 & 1.424064 \bullet 10^{-5} & 0.003195 & 0.001350 & 0.000961 & 0.000520 & 0.003112 & -0.000366 & 0.000825 \\
0.000616 & 0.004405 & -0.000397 & 0.005001 & 0.000961 & 0.002491 & 0.000216 & 0.004641 & -0.000825 & 0.001958 \\
0.001474 & -0.000181 & 0.001474 & 0.000181 & 0.000520 & 0.000216 & 0.001835 & 0.000000 & 0.000520 & -0.000216 \\
0.003686 & 0.017220 & -0.003686 & 0.017220 & 0.003112 & 0.004641 & 0.000000 & 0.017058 & -0.003112 & 0.004641 \\
1.424064 \bullet 10^{-5} & -0.003195 & 0.001187 & -0.003100 & -0.000366 & -0.000825 & 0.000520 & -0.003112 & 0.001350 & -0.000961 \\
0.000397 & 0.005001 & -0.000616 & 0.004405 & 0.000825 & 0.001958 & -0.000216 & 0.004641 & -0.000961 & 0.002491
\end{bmatrix}$$

So the complete inverted matrix equation is

$$\begin{bmatrix} u_2 \\ v_2 \\ u_3 \\ v_3 \\ u_5 \\ v_5 \\ u_6 \\ v_6 \\ u_7 \\ v_7 \end{bmatrix} = \begin{bmatrix}
0.003516 & 0.003956 & -0.00033 & 0.003734 & 0.001187 & 0.000616 & 0.001474 & 0.003686 & 0.000014 & 0.000397 \\
0.003956 & 0.018452 & -0.003734 & 0.016612 & 0.0031 & 0.004405 & -0.000181 & 0.01722 & -0.003195 & 0.005001 \\
-0.00033 & -0.003734 & 0.003516 & -0.003956 & 0.000014 & -0.000397 & 0.001474 & -0.003686 & 0.001187 & -0.000616 \\
0.003734 & 0.016612 & -0.003956 & 0.018452 & 0.003195 & 0.005001 & 0.000181 & 0.01722 & -0.0031 & 0.004405 \\
0.001187 & 0.0031 & 0.000014 & 0.003195 & 0.00135 & 0.000961 & 0.00052 & 0.003112 & -0.000366 & 0.000825 \\
0.000616 & 0.004405 & -0.000397 & 0.005001 & 0.000961 & 0.002491 & 0.000216 & 0.004641 & -0.000825 & 0.001958 \\
0.001474 & -0.000181 & 0.001474 & 0.000181 & 0.00052 & 0.000216 & 0.001835 & 0.0 & 0.00052 & -0.000216 \\
0.003686 & 0.01722 & -0.003686 & 0.01722 & 0.003112 & 0.004641 & 0.0 & 0.017058 & -0.003112 & 0.004641 \\
0.000014 & -0.003195 & 0.001187 & -0.0031 & -0.000366 & -0.000825 & 0.00052 & -0.003112 & 0.00135 & -0.000961 \\
0.000397 & 0.005001 & -0.000616 & 0.004405 & 0.000825 & 0.001958 & -0.000216 & 0.004641 & -0.000961 & 0.002491
\end{bmatrix} \begin{bmatrix} F_{u2} \\ F_{v2} \\ F_{u3} \\ F_{v3} \\ F_{u5} \\ F_{v5} \\ F_{u6} \\ F_{v6} \\ F_{u7} \\ F_{v7} \end{bmatrix}$$

Now we know all the terms in the force vector on the right hand side of the equation. They are

$$F := \begin{bmatrix} 1 \\ 0 \\ 1 \\ 0 \\ 0 \\ 0 \\ 0 \\ 0 \\ 0 \\ 0 \end{bmatrix}$$

So if we multiply F by R^{-1} we get the nodal displacements. The multiplication is simple because there are only two non-zero forces, so only columns one and three in the inverted matrix contribute to the value of each displacement. For example,

$$v_2 = (0.003956 \times 1) - (0.003734 \times 1) = 0.000222\,\text{m}$$

and

$$u_6 = (0.001474 \times 1) + 0(.001474 \times 1) = 0.002948\,\text{m}$$

146

The values of the eight nodal displacements (in metres) derived from the matrix product $R^{-1}F$ are:

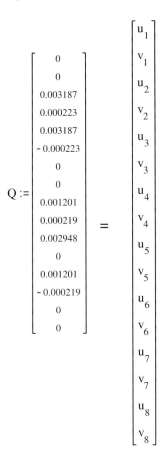

$$R^{-1} \cdot F = \begin{bmatrix} 0.003187 \\ 0.000223 \\ 0.003187 \\ -0.000223 \\ 0.001201 \\ 0.000219 \\ 0.002948 \\ 0.000000 \\ 0.001201 \\ -0.000219 \end{bmatrix} = \begin{bmatrix} u_2 \\ v_2 \\ u_3 \\ v_3 \\ u_5 \\ v_5 \\ u_6 \\ v_6 \\ u_7 \\ v_7 \end{bmatrix}$$

We can now form the complete set of nodal displacements by adding the known zero displacements u_1, v_1, u_4, v_4, u_8 and v_8. We call this complete nodal displacement vector Q.

$$Q := \begin{bmatrix} 0 \\ 0 \\ 0.003187 \\ 0.000223 \\ 0.003187 \\ -0.000223 \\ 0 \\ 0 \\ 0.001201 \\ 0.000219 \\ 0.002948 \\ 0 \\ 0.001201 \\ -0.000219 \\ 0 \\ 0 \end{bmatrix} = \begin{bmatrix} u_1 \\ v_1 \\ u_2 \\ v_2 \\ u_3 \\ v_3 \\ u_4 \\ v_4 \\ u_5 \\ v_5 \\ u_6 \\ v_6 \\ u_7 \\ v_7 \\ u_8 \\ v_8 \end{bmatrix}$$

8.1.8 ARE THE NODAL FORCES IN EQUILIBRIUM?

We are now in a position to calculate the stresses at the Gauss points and at the nodes. Before doing that, however it is instructive to find a check that will confirm that the nodal displacements we have calculated are reasonable. One such check is to calculate the applied nodal forces from the nodal displacement and see if this force system is in equilibrium, as it should be.

The stiffness matrix equation whose structure is defined in equation (8.1.2) relates the nodal forces to the nodal displacements through

$$F = KU.$$

We know all the 256 terms in the stiffness matrix K which are listed in eqn (8.1.1) and we have just calculated the 16 nodal displacements which are contained in the vector Q. So we have all we need to calculate the 16 force components F.

$$F = KQ.$$

This involves the multiplication of a 16×16 matrix and a 16×1 matrix which yields a 16×1 matrix – the list of the 16 force components. The multiplication process is tedious but the MATHCAD output for this matrix product is, where the forces are in MN,

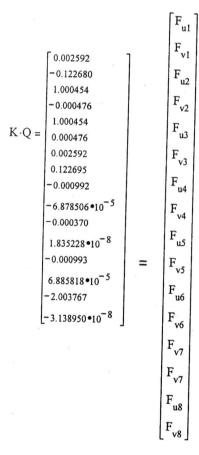

$$
K \cdot Q = \begin{bmatrix}
0.002592 \\
-0.122680 \\
1.000454 \\
-0.000476 \\
1.000454 \\
0.000476 \\
0.002592 \\
0.122695 \\
-0.000992 \\
-6.878506 \cdot 10^{-5} \\
-0.000370 \\
1.835228 \cdot 10^{-8} \\
-0.000993 \\
6.885818 \cdot 10^{-5} \\
-2.003767 \\
-3.138950 \cdot 10^{-8}
\end{bmatrix}
=
\begin{bmatrix}
F_{u1} \\
F_{v1} \\
F_{u2} \\
F_{v2} \\
F_{u3} \\
F_{v3} \\
F_{u4} \\
F_{v4} \\
F_{u5} \\
F_{v5} \\
F_{u6} \\
F_{v6} \\
F_{v7} \\
F_{v7} \\
F_{u8} \\
F_{v8}
\end{bmatrix}
$$

One condition of force equilibrium is that the resultant horizontal force on the element is zero, that is

$$F_{u1} + F_{u2} + F_{u3} + F_{u4} + F_{u5} + F_{u6} + F_{u7} + F_{u8} = 0$$

Taking the values from the calculated force vector:

$$0.002592 + 1.000454 + 1.000454 + 0.002592 - 0.000992 - 0.000370 - 0.000993 - 2.003767$$

$$= -0.00003MN$$

which is close enough to zero to confirm that this condition for static force equilibrium is satisfied and our confidence in the nodal displacement values is enhanced.

8.1.9 THE STRESSES AT THE GAUSS POINTS

In the last section of this example we will calculate the stresses at the 4 Gauss points and at the 8 nodes. We recall from equation (6.25) that the stress vector is the product of the elastic constants matrix and the strain matrix

$$\text{Stress} = D\varepsilon = DBu$$

In this example we have called the nodal displacement vector Q, so

$$\text{Stress} = DBQ$$

The 3×1 stress vector is

$$\begin{bmatrix} \sigma_{xx} \\ \sigma_{yy} \\ \sigma_{xy} \end{bmatrix}$$

We first calculate the stresses at the Gauss points. This is a particularly simple because we have already determined all the matrices in the matrix product that gives the stresses.

At Gauss point a the B matrix is Ba so the stresses are $D \cdot Ba \cdot Q$. We calculated the Ba matrix at the beginning of this exercise. Similarly the B matrix for gauss point b is Bb, for c is Bc and for d is Bd. We have all these matrices so the stresses, in MPa, follow:

GAUSS POINT STRESSES

$$D \cdot Ba \cdot Q = \begin{bmatrix} 77.55 \\ 10.49 \\ 1.92 \end{bmatrix} = \begin{bmatrix} \sigma_{xx} \\ \sigma_{yy} \\ \sigma_{xy} \end{bmatrix} \qquad D \cdot Bb \cdot Q = \begin{bmatrix} 140.61 \\ -6.05 \\ -13.83 \end{bmatrix} = \begin{bmatrix} \sigma_{xx} \\ \sigma_{yy} \\ \sigma_{xy} \end{bmatrix}$$

$$D \cdot Bc \cdot Q = \begin{bmatrix} 140.61 \\ -6.05 \\ 13.83 \end{bmatrix} = \begin{bmatrix} \sigma_{xx} \\ \sigma_{yy} \\ \sigma_{xy} \end{bmatrix} \qquad D \cdot Bd \cdot Q = \begin{bmatrix} 77.55 \\ 10.49 \\ -1.92 \end{bmatrix} = \begin{bmatrix} \sigma_{xx} \\ \sigma_{yy} \\ \sigma_{xy} \end{bmatrix}$$

It is evident that

(i) The direct stresses at a and d are equal as are those at b and c. This would be expected from the symmetry of the element.

(ii) The values of σ_{xx} at b and c are higher than those at a and d because of the lower area of cross-section of the element at b and c.

149

(iii) The value of the lateral stress σ_{yy} is tensile at a and d because of the lateral forces necessary to inhibit the lateral Poisson's ratio contraction near the fixed edge of the sheet. To balance this σ_{yy} changes to compression at Gauss points b and c.

(iv) Of all the element stresses that can be calculated those at Gauss points are the most accurate.

8.1.10 STRESSES AT THE NODES

The nodal stresses are calculated similarly using the matrix product relation

$$\text{stresses} = D \cdot B \cdot Q$$

Here, we recall B is the strain matrix which is the product of the two matrices A and G. We need to calculate these for the 8 nodes by substituting the master element nodal co-ordinates and the real element nodal co-ordinates in the constituent matrices A and G in exactly the same way that we did for the Gauss point matrices. It will be helpful to list the master element nodal co-ordinates:

NODE	ξ	η
1	-1	-1
2	$+1$	-1
3	$+1$	$+1$
4	-1	$+1$
5	0	-1
6	$+1$	0
7	0	$+1$
8	-1	0
Centroid	0	0

The detailed calculations that give the nodal stresses are given below.

Note that, at node 1, the A matrix is given the symbol $a1$ and the G matrix $g1$. The B matrix is $B1$ so

$$B1 = a1 \cdot g1 \qquad \text{and the stresses are given by } D \cdot B1 \cdot Q$$

Similarly for node 2 the corresponding matrices are $a2$, $g2$ and $B2$ and the stresses are given by $D \cdot B2 \cdot Q$, and so on for the other nodes.

NODE 1

$$a1 := 1.3333 \cdot \begin{bmatrix} .75 & -.25 & 0 & 0 \\ 0 & 0 & 0 & 1.0 \\ 0 & 1.0 & .75 & -.25 \end{bmatrix}$$

$$g1 := \begin{bmatrix} -1.5 & 0 & -.5 & 0 & 0 & 0 & 0 & 2 & 0 & 0 & 0 & 0 & 0 & 0 & 0 \\ -1.5 & 0 & 0 & 0 & 0 & 0 & -.5 & 0 & 0 & 0 & 0 & 0 & 0 & 0 & 2 & 0 \\ 0 & -1.5 & 0 & -.5 & 0 & 0 & 0 & 0 & 0 & 2 & 0 & 0 & 0 & 0 & 0 & 0 \\ 0 & -1.5 & 0 & 0 & 0 & 0 & 0 & -.5 & 0 & 0 & 0 & 0 & 0 & 0 & 0 & 2 \end{bmatrix}$$

$$a1 \cdot g1 = \begin{bmatrix} -1.000 & 0.000 & -0.500 & 0.000 & 0.000 & 0.000 & 0.167 & 0.000 & 2.000 & 0.000 & 0.000 & 0.000 & 0.000 & 0.000 & -0.667 & 0.000 \\ 0.000 & -2.000 & 0.000 & 0.000 & 0.000 & 0.000 & 0.000 & -0.667 & 0.000 & 0.000 & 0.000 & 0.000 & 0.000 & 0.000 & 0.000 & 2.667 \\ -2.000 & -1.000 & 0.000 & -0.500 & 0.000 & 0.000 & -0.667 & 0.167 & 0.000 & 2.000 & 0.000 & 0.000 & 0.000 & 0.000 & 2.667 & -0.667 \end{bmatrix}$$

$$B1 := a1 \cdot g1$$

$$D \cdot B1 \cdot Q = \begin{bmatrix} 61.57 \\ 18.47 \\ 8.70 \end{bmatrix} = \begin{bmatrix} \sigma_{xx} \\ \sigma_{yy} \\ \sigma_{xy} \end{bmatrix}$$

NODE 2

$$a2 := 4.0 \cdot \begin{bmatrix} .25 & -.25 & 0 & 0 \\ 0 & 0 & 0 & 1.0 \\ 0 & 1.0 & .25 & -.25 \end{bmatrix}$$

$$g2 := \begin{bmatrix} .5 & 0 & 1.5 & 0 & 0 & 0 & 0 & 0 & -2 & 0 & 0 & 0 & 0 & 0 & 0 & 0 \\ 0 & 0 & -1.5 & 0 & -.5 & 0 & 0 & 0 & 0 & 0 & 2 & 0 & 0 & 0 & 0 & 0 \\ 0 & .5 & 0 & 1.5 & 0 & 0 & 0 & 0 & 0 & -2 & 0 & 0 & 0 & 0 & 0 & 0 \\ 0 & 0 & 0 & -1.5 & 0 & -.5 & 0 & 0 & 0 & 0 & 2 & 0 & 0 & 0 & 0 & 0 \end{bmatrix}$$

$$a2 \cdot g2 = \begin{bmatrix} 0.500 & 0.000 & 3.000 & 0.000 & 0.500 & 0.000 & 0.000 & 0.000 & -2.000 & 0.000 & -2.000 & 0.000 & 0.000 & 0.000 & 0.000 & 0.000 \\ 0.000 & 0.000 & 0.000 & -6.000 & 0.000 & -2.000 & 0.000 & 0.000 & 0.000 & 0.000 & 0.000 & 8.000 & 0.000 & 0.000 & 0.000 & 0.000 \\ 0.000 & 0.500 & -6.000 & 3.000 & -2.000 & 0.500 & 0.000 & 0.000 & 0.000 & -2.000 & 8.000 & -2.000 & 0.000 & 0.000 & 0.000 & 0.000 \end{bmatrix}$$

$$B2 := a2 \cdot g2$$

$$D \cdot B2 \cdot Q = \begin{bmatrix} 197.14 \\ -2.67 \\ -47.77 \end{bmatrix} = \begin{bmatrix} \sigma_{xx} \\ \sigma_{yy} \\ \sigma_{xy} \end{bmatrix}$$

NODE 3

$$a3 := 4.0 \cdot \begin{bmatrix} .25 & .25 & 0 & 0 \\ 0 & 0 & 0 & 1.0 \\ 0 & 1.0 & .25 & .25 \end{bmatrix}$$

$$g3 := \begin{bmatrix} 0 & 0 & 0 & 0 & 1.5 & 0 & .5 & 0 & 0 & 0 & 0 & 0 & -2 & 0 & 0 & 0 \\ 0 & 0 & .5 & 0 & 1.5 & 0 & 0 & 0 & 0 & 0 & -2.0 & 0 & 0 & 0 & 0 & 0 \\ 0 & 0 & 0 & 0 & 0 & 1.5 & 0 & .5 & 0 & 0 & 0 & 0 & 0 & -2 & 0 & 0 \\ 0 & 0 & 0 & .5 & 0 & 1.5 & 0 & 0 & 0 & 0 & 0 & -2.0 & 0 & 0 & 0 & 0 \end{bmatrix}$$

$$a3 \cdot g3 = \begin{bmatrix} 0.000 & 0.000 & 0.500 & 0.000 & 3.000 & 0.000 & 0.500 & 0.000 & 0.000 & 0.000 & -2.000 & 0.000 & -2.000 & 0.000 & 0.000 & 0.000 \\ 0.000 & 0.000 & 0.000 & 2.000 & 0.000 & 6.000 & 0.000 & 0.000 & 0.000 & 0.000 & 0.000 & -8.000 & 0.000 & 0.000 & 0.000 & 0.000 \\ 0.000 & 0.000 & 2.000 & 0.500 & 6.000 & 3.000 & 0.000 & 0.500 & 0.000 & 0.000 & -8.000 & -2.000 & 0.000 & -2.000 & 0.000 & 0.000 \end{bmatrix}$$

$$B3 := a3 \cdot g3$$

$$D \cdot B3 \cdot Q = \begin{bmatrix} 197.14 \\ -2.67 \\ 47.77 \end{bmatrix} \qquad = \qquad \begin{bmatrix} \sigma_{xx} \\ \sigma_{yy} \\ \sigma_{xy} \end{bmatrix}$$

NODE 4

$$a4 := 1.3333 \cdot \begin{bmatrix} .75 & .25 & 0 & 0 \\ 0 & 0 & 0 & 1 \\ 0 & 1 & .75 & .25 \end{bmatrix}$$

$$g4 := \begin{bmatrix} 0 & 0 & 0 & 0 & -.5 & 0 & -1.5 & 0 & 0 & 0 & 0 & 0 & 2 & 0 & 0 & 0 \\ .5 & 0 & 0 & 0 & 0 & 0 & 1.5 & 0 & 0 & 0 & 0 & 0 & 0 & 0 & -2 & 0 \\ 0 & 0 & 0 & 0 & 0 & -.5 & 0 & -1.5 & 0 & 0 & 0 & 0 & 0 & 2 & 0 & 0 \\ 0 & .5 & 0 & 0 & 0 & 0 & 0 & 1.5 & 0 & 0 & 0 & 0 & 0 & 0 & 0 & -2 \end{bmatrix}$$

$$a4 \cdot g4 = \begin{bmatrix} 0.167 & 0.000 & 0.000 & 0.000 & -0.500 & 0.000 & -1.000 & 0.000 & 0.000 & 0.000 & 0.000 & 0.000 & 2.000 & 0.000 & -0.667 & 0.000 \\ 0.000 & 0.667 & 0.000 & 0.000 & 0.000 & 0.000 & 0.000 & 2.000 & 0.000 & 0.000 & 0.000 & 0.000 & 0.000 & 0.000 & 0.000 & -2.667 \\ 0.667 & 0.167 & 0.000 & 0.000 & 0.000 & -0.500 & 2.000 & -1.000 & 0.000 & 0.000 & 0.000 & 0.000 & 2.000 & -2.667 & -0.667 \end{bmatrix}$$

$$B4 := a4 \cdot g4$$

$$D \cdot B4 \cdot Q = \begin{bmatrix} 61.57 \\ 18.47 \\ -8.70 \end{bmatrix} \qquad = \qquad \begin{bmatrix} \sigma_{xx} \\ \sigma_{yy} \\ \sigma_{xy} \end{bmatrix}$$

152

NODE 5

$$a5 := 2 \cdot \begin{bmatrix} .5 & -.25 & 0 & 0 \\ 0 & 0 & 0 & 1.0 \\ 0 & 1.0 & .5 & -.25 \end{bmatrix}$$

$$g5 := \begin{bmatrix} -.5 & 0 & .5 & 0 & 0 & 0 & 0 & 0 & 0 & 0 & 0 & 0 & 0 & 0 & 0 & 0 \\ -.5 & 0 & -.5 & 0 & -.5 & 0 & -.5 & 0 & -.5 & 0 & 1.0 & 0 & .5 & 0 & 1.0 & 0 \\ 0 & -.5 & 0 & .5 & 0 & 0 & 0 & 0 & 0 & 0 & 0 & 0 & 0 & 0 & 0 & 0 \\ 0 & -.5 & 0 & -.5 & 0 & -.5 & 0 & -.5 & -0 & -.5 & 0 & 1.0 & 0 & .5 & 0 & 1.0 \end{bmatrix}$$

$$a5 \cdot g5 = \begin{bmatrix} -0.250 & 0.000 & 0.750 & 0.000 & 0.250 & 0.000 & 0.250 & 0.000 & 0.250 & 0.000 & -0.500 & 0.000 & -0.250 & 0.000 & -0.500 & 0.000 \\ 0.000 & -1.000 & 0.000 & -1.000 & 0.000 & -1.000 & 0.000 & -1.000 & 0.000 & -1.000 & 0.000 & 2.000 & 0.000 & 1.000 & 0.000 & 2.000 \\ -1.000 & -0.250 & -1.000 & 0.750 & -1.000 & 0.250 & -1.000 & 0.250 & -1.000 & 0.250 & 2.000 & -0.500 & 1.000 & -0.250 & 2.000 & -0.500 \end{bmatrix}$$

$$B5 := a5 \cdot g5$$

$$D \cdot B5 \cdot Q = \begin{bmatrix} 120.44 \\ 5.78 \\ -6.85 \end{bmatrix} = \begin{bmatrix} \sigma_{xx} \\ \sigma_{yy} \\ \sigma_{xy} \end{bmatrix}$$

NODE 6

$$a6 := 4 \cdot \begin{bmatrix} .25 & 0 & 0 & 0 \\ 0 & 0 & 0 & 1.0 \\ 0 & 1.0 & .25 & 0 \end{bmatrix}$$

$$g6 := \begin{bmatrix} .5 & 0 & .5 & 0 & .5 & 0 & .5 & 0 & -1 & 0 & .5 & 0 & -1 & 0 & -.5 & 0 \\ 0 & 0 & -.5 & 0 & .5 & 0 & 0 & 0 & 0 & 0 & 0 & 0 & 0 & 0 & 0 & 0 \\ 0 & .5 & 0 & .5 & 0 & .5 & 0 & .5 & 0 & -1 & 0 & .5 & 0 & -1 & 0 & -.5 \\ 0 & 0 & 0 & -.5 & 0 & .5 & 0 & 0 & 0 & 0 & 0 & 0 & 0 & 0 & 0 & 0 \end{bmatrix}$$

$$a6 \cdot g6 = \begin{bmatrix} 0.500 & 0.000 & 0.500 & 0.000 & 0.500 & 0.000 & 0.500 & 0.000 & -1.000 & 0.000 & 0.500 & 0.000 & -1.000 & 0.000 & -0.500 & 0.000 \\ 0.000 & 0.000 & 0.000 & -2.000 & 0.000 & 2.000 & 0.000 & 0.000 & 0.000 & 0.000 & 0.000 & 0.000 & 0.000 & 0.000 & 0.000 & 0.000 \\ 0.000 & 0.500 & -2.000 & 0.500 & 2.000 & 0.500 & 0.000 & 0.500 & 0.000 & -1.000 & 0.000 & 0.500 & 0.000 & -1.000 & 0.000 & -0.500 \end{bmatrix}$$

$$B6 := a6 \cdot g6$$

$$D \cdot B6 \cdot Q = \begin{bmatrix} 151.65 \\ -16.32 \\ 0.00 \end{bmatrix} = \begin{bmatrix} \sigma_{xx} \\ \sigma_{yy} \\ \sigma_{xy} \end{bmatrix}$$

NODE 7

$$a7 := 2 \cdot \begin{bmatrix} .5 & .25 & 0 & 0 \\ 0 & 0 & 0 & 1 \\ 0 & 1 & .5 & .25 \end{bmatrix}$$

$$g7 := \begin{bmatrix} 0 & 0 & 0 & 0 & .5 & 0 & -.5 & 0 & 0 & 0 & 0 & 0 & 0 & 0 & 0 & 0 \\ .5 & 0 & .5 & 0 & .5 & 0 & .5 & 0 & -.5 & 0 & -1 & 0 & .5 & 0 & -1 & 0 \\ 0 & 0 & 0 & 0 & 0 & .5 & 0 & -.5 & 0 & 0 & 0 & 0 & 0 & 0 & 0 & 0 \\ 0 & .5 & 0 & .5 & 0 & .5 & 0 & .5 & 0 & -.5 & 0 & -1 & 0 & .5 & 0 & -1 \end{bmatrix}$$

$$a7 \cdot g7 = \begin{bmatrix} 0.250 & 0.000 & 0.250 & 0.000 & 0.750 & 0.000 & -0.250 & 0.000 & -0.250 & 0.000 & -0.500 & 0.000 & 0.250 & 0.000 & -0.500 & 0.000 \\ 0.000 & 1.000 & 0.000 & 1.000 & 0.000 & 1.000 & 0.000 & 1.000 & 0.000 & -1.000 & 0.000 & -2.000 & 0.000 & 1.000 & 0.000 & -2.000 \\ 1.000 & 0.250 & 1.000 & 0.250 & 1.000 & 0.750 & 1.000 & -0.250 & -1.000 & -0.250 & -2.000 & -0.500 & 1.000 & 0.250 & -2.000 & -0.500 \end{bmatrix}$$

$$B7 := a7 \cdot g7$$

$$D \cdot B7 \cdot Q = \begin{bmatrix} 120.44 \\ 5.78 \\ 6.85 \end{bmatrix} \quad = \quad \begin{bmatrix} \sigma_{xx} \\ \sigma_{yy} \\ \sigma_{xy} \end{bmatrix}$$

NODE 8

$$a8 := 1.3333 \cdot \begin{bmatrix} .75 & 0 & 0 & 0 \\ 0 & 0 & 0 & 1 \\ 0 & 1 & .75 & 0 \end{bmatrix}$$

$$g8 := \begin{bmatrix} -.5 & 0 & -.5 & 0 & -.5 & 0 & -.5 & 0 & 1 & 0 & .5 & 0 & 1 & 0 & -.5 & 0 \\ -.5 & 0 & 0 & 0 & 0 & 0 & .5 & 0 & 0 & 0 & 0 & 0 & 0 & 0 & 0 & 0 \\ 0 & -.5 & 0 & -.5 & 0 & -.5 & 0 & -.5 & 0 & 1 & 0 & .5 & 0 & 1 & 0 & -.5 \\ 0 & -.5 & 0 & 0 & 0 & 0 & 0 & .5 & 0 & 0 & 0 & 0 & 0 & 0 & 0 & 0 \end{bmatrix}$$

$$a8 \cdot g8 = \begin{bmatrix} -0.500 & 0.000 & -0.500 & 0.000 & -0.500 & 0.000 & -0.500 & 0.000 & 1.000 & 0.000 & 0.500 & 0.000 & 1.000 & 0.000 & -0.500 & 0.000 \\ 0.000 & -0.667 & 0.000 & 0.000 & 0.000 & 0.000 & 0.000 & 0.667 & 0.000 & 0.000 & 0.000 & 0.000 & 0.000 & 0.000 & 0.000 & 0.000 \\ -0.667 & -0.500 & 0.000 & -0.500 & 0.000 & -0.500 & 0.667 & -0.500 & 0.000 & 1.000 & 0.000 & 0.500 & 0.000 & 1.000 & 0.000 & -0.500 \end{bmatrix}$$

$$B8 := a8 \cdot g8$$

$$D \cdot B8 \cdot Q = \begin{bmatrix} 52.47 \\ 15.74 \\ 0.00 \end{bmatrix} \quad = \quad \begin{bmatrix} \sigma_{xx} \\ \sigma_{yy} \\ \sigma_{xy} \end{bmatrix}$$

CENTROID

This is located at the centre of the element at $\xi=0$ and $\eta=0$

$$a9 := 2.0 \cdot \begin{bmatrix} .5 & 0 & 0 & 0 \\ 0 & 0 & 0 & 1 \\ 0 & 1 & .5 & 0 \end{bmatrix}$$

$$g9 := \begin{bmatrix} 0 & 0 & 0 & 0 & 0 & 0 & 0 & 0 & 0 & 0 & .5 & 0 & 0 & 0 & -.5 & 0 \\ 0 & 0 & 0 & 0 & 0 & 0 & 0 & 0 & -.5 & 0 & 0 & 0 & .5 & 0 & 0 & 0 \\ 0 & 0 & 0 & 0 & 0 & 0 & 0 & 0 & 0 & 0 & 0 & .5 & 0 & 0 & 0 & -.5 \\ 0 & 0 & 0 & 0 & 0 & 0 & 0 & 0 & 0 & -.5 & 0 & 0 & 0 & .5 & 0 & 0 \end{bmatrix}$$

$$a9 \cdot g9 = \begin{bmatrix} 0.000 & 0.000 & 0.000 & 0.000 & 0.000 & 0.000 & 0.000 & 0.000 & 0.000 & 0.000 & 0.500 & 0.000 & 0.000 & 0.000 & -0.500 & 0.000 \\ 0.000 & 0.000 & 0.000 & 0.000 & 0.000 & 0.000 & 0.000 & 0.000 & 0.000 & -1.000 & 0.000 & 0.000 & 0.000 & 1.000 & 0.000 & 0.000 \\ 0.000 & 0.000 & 0.000 & 0.000 & 0.000 & 0.000 & 0.000 & 0.000 & -1.000 & 0.000 & 0.000 & 0.500 & 1.000 & 0.000 & 0.000 & -0.500 \end{bmatrix}$$

$$B9 := a9 \cdot g9$$

$$D \cdot B9 \cdot Q = \begin{bmatrix} 102.24 \\ 0.32 \\ 0.00 \end{bmatrix} \qquad = \qquad \begin{bmatrix} \sigma_{xx} \\ \sigma_{yy} \\ \sigma_{xy} \end{bmatrix}$$

The nodal direct stresses σ_{xx} and σ_{yy} are collected together in Fig. 8.2. We shall need to comment on them in some detail later, but we note at this stage that the values of σ_{xx} fall into an expected range. At node 2 the stress is 197.14 MPa. The average stress at the free end of the plate is the

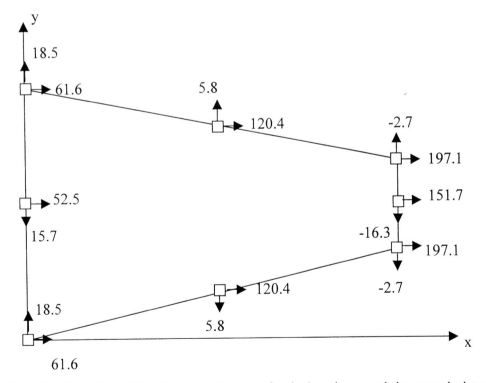

Fig. 8.2 The values of the direct nodal stresses for the 8-node tapered element calculated using four integration points.

total load, 2 MN, divided by the area of cross-section $0.02 \times 0.5 = 0.01\,\mathrm{m}^2$, that is 200 MPa, which is close to the node 2 value. The corresponding stress at node 1 is 61.57 MPa and the average stress at the fixed end is $2/(1.5 \times 0.02) = 66.66$ MPa which is close to the node 1 value.

We work through this example again in the next section again using eight nodes but increasing the number of integration points to nine. This gives further insight into the Gauss quadrature numerical integration process and also gives an indication of the effect of this more detailed integration procedure on the calculated stresses. We shall see that the volume of computation in this second version is much greater than was the case with the four Gauss points we have just used. So we shall ask the question 'was it worth it?'

8.2 A REPEAT OF THE ANALYSIS OF THE EIGHT-NODE QUADRILATERAL TAPERED ELEMENT IN PLANE STRESS BUT USING NINE GAUSS INTEGRATION POINTS

The analysis follows the same structure as that for the previous case with four integration points so this presentation will not need to be annotated in such detail as was the previous example.

8.2.1 SIMILARITIES TO AND DIFFERENCE FROM THE PREVIOUS CASE

The principal differences from the previous case are:

(i) Since there are nine Gauss points, the first section of the analysis will involve the derivation of nine contributions to the element stiffness matrix, in contrast to four for the previous case.
    The strain matrix $B = AG$ is determined by the number of nodes which is the same in this example as for the previous case with four integration points so the expressions for the A and G matrices and the shape functions are the same in both cases. However, the value of the terms in the matrices will differ because the ξ, η co-ordinates of the nine Gauss points are different from those of the four Gauss points.
(ii) The weights at the Gauss points, which are used in the numerical integration of the triple matrix product, are different. The nine Gauss point weights have been derived in Section 7.4.2 and are presented in Fig. 7.24.

The real element with the nodes and the Gauss points identified is drawn in Fig. 8.3 together with the master element from which it is derived.

We list below the shape function derivatives for the 8-node element, the nodal coordinates in the real element and the Jacobians. These are all identical to those in the four-integration-point case but we reproduce them for ease of reference.

The new information listed are the (ξ, η) coordinates of the nine Gauss points in the master element, which of course differ from those of the four Gauss points.

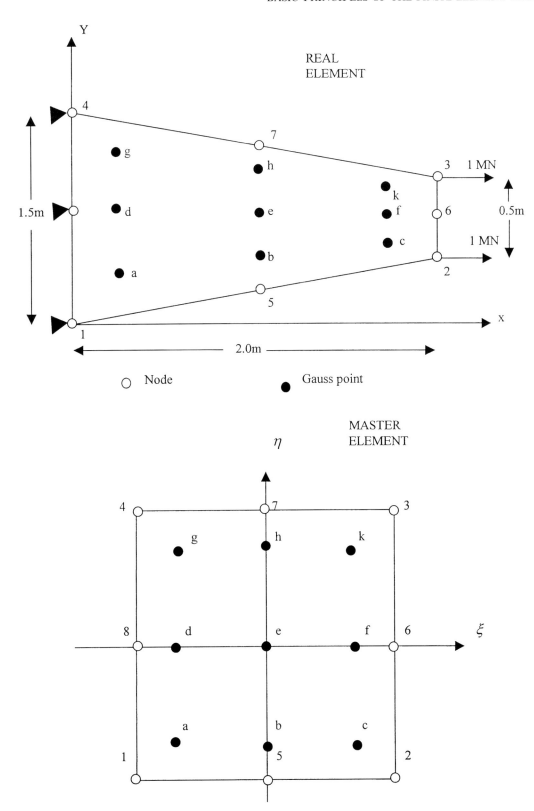

Fig. 8.3   The Real Tapered Element with nine Gauss points and its Master Element

Shape function derivatives for 8-node serendipity element

Xi derivatives　　　　　　eta derivatives

$$\xi := -0.7746 \qquad \eta := 0$$

$s1 := 0.25 \cdot (1 - \eta) \cdot (2 \cdot \xi + \eta)$

$\qquad s1 = -0.38730$

$s2 := 0.25 \cdot (1 - \eta) \cdot (2 \cdot \xi - \eta)$

$\qquad s2 = -0.38730$

$s3 := 0.25 \cdot (1 + \eta) \cdot (2 \cdot \xi + \eta)$

$\qquad s3 = -0.38730$

$s4 := 0.25 \cdot (1 + \eta) \cdot (2 \cdot \xi - \eta)$

$\qquad s4 = -0.38730$

$s5 := -\xi \cdot (1 - \eta)$

$\qquad s5 = 0.77460$

$s6 := 0.5 \cdot \left(1 - \eta^2\right)$

$\qquad s6 = 0.50000$

$s7 := -\xi \cdot (1 + \eta)$

$\qquad s7 = 0.77460$

$s8 := -0.5 \cdot \left(1 - \eta^2\right)$

$\qquad s8 = -0.50000$

$t1 := 0.25 \cdot (1 - \xi) \cdot (2 \cdot \eta + \xi)$

$\qquad t1 = -0.34365$

$t2 := -0.25 \cdot (1 + \xi) \cdot (\xi - 2 \cdot \eta)$

$\qquad t2 = 0.04365$

$t3 := 0.25 \cdot (1 + \xi) \cdot (2 \cdot \eta + \xi)$

$\qquad t3 = -0.04365$

$t4 := 0.25 \cdot (1 - \xi) \cdot (2 \cdot \eta - \xi)$

$\qquad t4 = 0.34365$

$t5 := -0.5 \cdot \left(1 - \xi^2\right)$

$\qquad t5 = -0.20000$

$t6 := -\eta \cdot (1 + \xi)$

$\qquad t6 = 0.00000$

$t7 := 0.5 \cdot \left(1 - \xi^2\right)$

$\qquad t7 = 0.20000$

$t8 := -\eta \cdot (1 - \xi)$

$\qquad t8 = 0.00000$

$$G := \begin{bmatrix} s1 & 0 & s2 & 0 & s3 & 0 & s4 & 0 & s5 & 0 & s6 & 0 & s7 & 0 & s8 & 0 \\ t1 & 0 & t2 & 0 & t3 & 0 & t4 & 0 & t5 & 0 & t6 & 0 & t7 & 0 & t8 & 0 \\ 0 & s1 & 0 & s2 & 0 & s3 & 0 & s4 & 0 & s5 & 0 & s6 & 0 & s7 & 0 & s8 \\ 0 & t1 & 0 & t2 & 0 & t3 & 0 & t4 & 0 & t5 & 0 & t6 & 0 & t7 & 0 & t8 \end{bmatrix}$$

Nodal coordinates of the real element

$x1 := 0 \qquad y1 := 0$

$x2 := 2 \qquad y2 := .5$

$x3 := 2 \qquad y3 := 1$

$x4 := 0 \qquad y4 := 1.5$

$x5 := 1 \qquad y5 := .25$

$x6 := 2 \qquad y6 := .75$

$x7 := 1 \qquad y7 := 1.25$

$x8 := 0 \qquad y8 := .75$

Expressions for the Jacobians

$$J_{11} := s1 \cdot x1 + s2 \cdot x2 + s3 \cdot x3 + s4 \cdot x4 + s5 \cdot x5 + s6 \cdot x6 + s7 \cdot x7 + s8 \cdot x8$$

$$J_{11} = 1.000000$$

$$J_{12} := s1 \cdot y1 + s2 \cdot y2 + s3 \cdot y3 + s4 \cdot y4 + s5 \cdot y5 + s6 \cdot y6 + s7 \cdot y7 + s8 \cdot y8$$

$$J_{12} = 0.000000$$

$$J_{21} := t1 \cdot x1 + t2 \cdot x2 + t3 \cdot x3 + t4 \cdot x4 + t5 \cdot x5 + t6 \cdot x6 + t7 \cdot x7 + t8 \cdot x8$$

$$J_{21} = 0.000000$$

$$J_{22} := t1 \cdot y1 + t2 \cdot y2 + t3 \cdot y3 + t4 \cdot y4 + t5 \cdot y5 + t6 \cdot y6 + t7 \cdot y7 + t8 \cdot y8$$

$$J_{22} = 0.693650$$

$$detJ := J_{11} \cdot J_{22} - J_{21} \cdot J_{12}$$

$$detJ = 0.69365$$

$$A := \frac{1}{detJ} \begin{bmatrix} J_{22} & -J_{12} & 0 & 0 \\ 0 & 0 & -J_{21} & J_{11} \\ -J_{21} & J_{11} & J_{22} & -J_{12} \end{bmatrix}$$

Coordinates of the integration points in the Master Element

Gauss point	ξ	η
a	−0.7746	−0.7746
b	0	−0.7746
c	0.7746	−0.7746
d	−0.7746	0
e	0	0
f	0.7746	0
g	−0.7746	0.7746
h	0	0.7746
k	0.7746	0.7746

8.2.2 CALCULATION OF THE STIFFNESS MATRIX B AND THE MATRIX PRODUCT $B^T \cdot D \cdot B \cdot \det J \cdot t$ AT THE NINE INTEGRATION POINTS

There follows the calculation of the stiffness matrix of the element. This involves calculating the matrix product

$$B^T \cdot D \cdot B \cdot \det J \cdot t$$

at each of the nine integration points, multiplying each by the appropriate Gauss point weight and summing these product to get the element stiffness matrix.

159

The matrix B is the product of the two matrices A and G which are defined in the previous example.

The calculation of the nine matrix products follows. For Gauss point a the matrices are Aa, Ga and Ba. The matrix product is Ka. For Gauss point b they are Ab, Gb, Bb and Kb – and so on.

CALCULATION OF THE MATRIX PRODUCT $B^T \cdot D \cdot B \cdot \det J \cdot t$ FOR THE NINE GAUSS POINTS

Gauss Point a

$$Aa := \begin{bmatrix} .69365 & -.19365 & 0 & 0 \\ 0 & 0 & 0 & 1 \\ 0 & 1 & .69365 & -.19365 \end{bmatrix} \cdot \frac{1}{.69365}$$

$$Ga := \begin{bmatrix} -1.0309 & 0.0 & -0.3436 & 0.0 & -0.1309 & 0.0 & -0.0436 & 0.0 & 1.3746 & 0.0 & 0.2 & 0.0 & 0.1746 & 0.0 & -0.2 & 0.0 \\ -1.0309 & 0.0 & -0.0436 & 0.0 & -0.1309 & 0.0 & -0.3436 & 0.0 & -0.2 & 0.0 & 0.1746 & 0.0 & 0.2 & 0.0 & 1.3746 & 0.0 \\ 0.0 & -1.0309 & 0.0 & -0.3436 & 0.0 & -0.1309 & 0.0 & -0.0436 & 0.0 & 1.3746 & 0.0 & 0.2 & 0.0 & 0.1746 & 0.0 & -0.2 \\ 0.0 & -1.0309 & 0.0 & -0.0436 & 0.0 & -0.1309 & 0.0 & -0.3436 & 0.0 & -0.2 & 0.0 & 0.1746 & 0.0 & 0.2 & 0.0 & 1.3746 \end{bmatrix}$$

$$Ba := Aa \cdot Ga$$

$$Ba = \begin{bmatrix} -0.7431 & 0.0000 & -0.3314 & 0.0000 & -0.0944 & 0.0000 & 0.0523 & 0.0000 & 1.4304 & 0.0000 & 0.1513 & 0.0000 & 0.1188 & 0.0000 & -0.5838 & 0.0000 \\ 0.0000 & -1.4862 & 0.0000 & -0.0629 & 0.0000 & -0.1887 & 0.0000 & -0.4954 & 0.0000 & -0.2883 & 0.0000 & 0.2517 & 0.0000 & 0.2883 & 0.0000 & 1.9817 \\ -1.4862 & -0.7431 & -0.0629 & -0.3314 & -0.1887 & -0.0944 & -0.4954 & 0.0523 & -0.2883 & 1.4304 & 0.2517 & 0.1513 & 0.2883 & 0.1188 & 1.9817 & -0.5838 \end{bmatrix}$$

For computational purposes it is not necessary to display the B matrix so we shall not do so for the remaining Gauss points. We remind ourselves of the elastic constants matrix by copying it below.

$$D := \begin{bmatrix} 76150 & 22845 & 0 \\ 22845 & 76150 & 0 \\ 0 & 0 & 26652 \end{bmatrix}$$

$$Ka := Ba^T \cdot D \cdot Ba \cdot .69365 \cdot .02$$

$$Ka = \begin{bmatrix}
1400.0 & 758.4 & 294.7 & 196.9 & 177.8 & 96.3 & 231.1 & 87.9 & -964.5 & -718.1 & -257.1 & -142.4 & -251.7 & -133.2 & -630.7 & -145.9 \\
758.4 & 2537.6 & 173.4 & 189.7 & 96.3 & 322.2 & 111.5 & 763.4 & -594.5 & 59.7 & -140.4 & -436.8 & -135.2 & -485.3 & -269.5 & -2951.0 \\
294.7 & 173.4 & 117.5 & 14.3 & 37.4 & 22.0 & -6.8 & 50.8 & -494.1 & -3.0 & -58.8 & -30.0 & -48.3 & -33.0 & 158.3 & -194.6 \\
196.9 & 189.7 & 14.3 & 44.8 & 25.0 & 24.1 & 59.7 & 26.5 & 6.8 & -156.1 & -33.9 & -35.2 & -37.7 & -33.7 & -231.2 & -60.1 \\
177.8 & 96.3 & 37.4 & 25.0 & 22.6 & 12.2 & 29.3 & 11.2 & -122.5 & -91.2 & -32.6 & -18.1 & -32.0 & -16.9 & -80.1 & -18.5 \\
96.3 & 322.2 & 22.0 & 24.1 & 12.2 & 40.9 & 14.2 & 96.9 & -75.5 & 7.6 & -17.8 & -55.5 & -17.2 & -61.6 & -34.2 & -374.7 \\
231.1 & 111.5 & -6.8 & 59.7 & 29.3 & 14.2 & 93.6 & -17.8 & 131.9 & -266.8 & -37.7 & -23.5 & -46.2 & -17.0 & -395.2 & 139.8 \\
87.9 & 763.4 & 50.8 & 26.5 & 11.2 & 96.9 & -17.8 & 260.2 & -230.1 & 178.6 & -18.9 & -128.8 & -13.1 & -148.6 & 130.0 & -1048.3 \\
-964.5 & -594.5 & -494.1 & 6.8 & -122.5 & -75.5 & 131.9 & -230.1 & 2192.3 & -283.2 & 201.7 & 98.0 & 148.7 & 118.1 & -1093.4 & 960.6 \\
-718.1 & 59.7 & -3.0 & -156.1 & -91.2 & 7.6 & -266.8 & 178.6 & -283.2 & 844.4 & 119.3 & 3.3 & 141.6 & -25.0 & 1101.4 & -912.4 \\
-257.1 & -140.4 & -58.8 & -33.9 & -32.6 & -17.8 & -37.7 & -18.9 & 201.7 & 119.3 & 47.6 & 26.1 & 45.8 & 24.9 & 91.2 & 40.7 \\
-142.4 & -436.8 & -30.0 & -35.2 & -18.1 & -55.5 & -23.5 & -128.8 & 98.0 & 3.3 & 26.1 & 75.4 & 25.6 & 83.3 & 64.3 & 494.3 \\
-251.7 & -135.2 & -48.3 & -37.7 & -32.0 & -17.2 & -46.2 & -13.1 & 148.7 & 141.6 & 45.8 & 25.6 & 45.6 & 23.5 & 138.0 & 12.4 \\
-133.2 & -485.3 & -33.0 & -33.7 & -16.9 & -61.6 & -17.0 & -148.6 & 118.1 & -25.0 & 24.9 & 83.3 & 23.5 & 93.0 & 33.7 & 578.0 \\
-630.7 & -269.5 & 158.3 & -231.2 & -80.1 & -34.2 & -395.2 & 130.0 & -1093.4 & 1101.4 & 91.2 & 64.3 & 138.0 & 33.7 & 1812.0 & -794.4 \\
-145.9 & -2951.0 & -194.6 & -60.1 & -18.5 & -374.7 & 139.8 & -1048.3 & 960.6 & -912.4 & 40.7 & 494.3 & 12.4 & 578.0 & -794.4 & 4274.7
\end{bmatrix}$$

Gauss Point b

$$Ab := \begin{bmatrix}
.5 & -.19365 & 0 & 0 \\
0 & 0 & 0 & 1 \\
0 & 1 & .5 & -.19365
\end{bmatrix} \cdot \frac{1}{.5}$$

$$Gb := \begin{bmatrix}
-0.34365 & 0.0 & 0.34365 & 0.0 & -0.04365 & 0.0 & 0.04365 & 0.0 & 0.0 & 0.0 & 0.2 & 0.0 & 0.0 & 0.0 & -0.2 & 0.0 \\
-0.3873 & 0.0 & -0.3873 & 0.0 & -0.3873 & 0.0 & -0.3873 & 0.0 & -0.5 & 0.0 & 0.7746 & 0.0 & 0.5 & 0.0 & 0.7746 & 0.0 \\
0.0 & -0.34365 & 0.0 & 0.34365 & 0.0 & -0.04365 & 0.0 & 0.04365 & 0.0 & 0.0 & 0.0 & 0.2 & 0.0 & 0.0 & 0.0 & -0.2 \\
0.0 & -0.3873 & 0.0 & -0.3873 & 0.0 & -0.3873 & 0.0 & -0.3873 & 0.0 & -0.5 & 0.0 & 0.7746 & 0.0 & 0.5 & 0.0 & 0.7746
\end{bmatrix}$$

$$Bb := Ab \cdot Gb$$

$$Kb := Bb^T \cdot D \cdot Bb \cdot 0.5 \cdot .02$$

$$Kb = \begin{bmatrix}
188.5 & 74.2 & 87.1 & -67.6 & 144.2 & 12.3 & 131.4 & -5.7 & 177.9 & 4.3 & -305.1 & -47.9 & -177.9 & -4.3 & -246.1 & 34.7 \\
74.2 & 466.9 & -47.4 & 431.4 & 21.2 & 451.4 & 5.7 & 446.9 & 17.3 & 579.9 & -62.3 & -908.6 & -17.3 & -579.9 & 8.5 & -888.0 \\
87.1 & -47.4 & 345.5 & -189.3 & 199.9 & -109.3 & 232.7 & -127.3 & 279.2 & -152.8 & -357.4 & 195.4 & -279.2 & 152.8 & -507.8 & 277.9 \\
-67.6 & 431.4 & -189.3 & 521.9 & -120.7 & 470.9 & -136.2 & 482.4 & -165.8 & 615.3 & 221.5 & -927.0 & 165.8 & -615.3 & 292.3 & -979.6 \\
144.2 & 21.2 & 199.9 & -120.7 & 168.5 & -40.8 & 175.6 & -58.8 & 222.1 & -64.3 & -327.9 & 58.3 & -222.1 & 64.3 & -360.3 & 140.9 \\
12.3 & 451.4 & -109.3 & 470.9 & -40.8 & 459.9 & -56.2 & 462.4 & -62.6 & 595.3 & 61.6 & -916.6 & 62.6 & -595.3 & 132.4 & -928.0 \\
131.4 & 5.7 & 232.7 & -136.2 & 175.6 & -56.2 & 188.5 & -74.2 & 235.0 & -84.2 & -334.6 & 89.2 & -235.0 & 84.2 & -393.6 & 171.8 \\
-5.7 & 446.9 & -127.3 & 482.4 & -58.8 & 462.4 & -74.2 & 466.9 & -85.9 & 599.9 & 97.7 & -919.0 & 85.9 & -599.9 & 168.4 & -939.6 \\
177.9 & 17.3 & 279.2 & -165.8 & 222.1 & -62.6 & 235.0 & -85.9 & 295.1 & -95.9 & -427.6 & 95.2 & -295.1 & 95.9 & -486.6 & 201.8 \\
4.3 & 579.9 & -152.8 & 615.3 & -64.3 & 595.3 & -84.2 & 599.9 & -95.9 & 771.5 & 102.8 & -1.2{\bullet}10^3 & 95.9 & -771.5 & 194.2 & -1.2{\bullet}10^3 \\
-305.1 & -62.3 & -357.4 & 221.5 & -327.9 & 61.6 & -334.6 & 97.7 & -427.6 & 102.8 & 647.3 & -76.7 & 427.6 & -102.8 & 677.7 & -241.8 \\
-47.9 & -908.6 & 195.4 & -927.0 & 58.3 & -916.6 & 89.2 & -919.0 & 95.2 & -1.2{\bullet}10^3 & -76.7 & 1.8{\bullet}10^3 & -95.2 & 1.2{\bullet}10^3 & -218.2 & 1.8{\bullet}10^3 \\
-177.9 & -17.3 & -279.2 & 165.8 & -222.1 & 62.6 & -235.0 & 85.9 & -295.1 & 95.9 & 427.6 & -95.2 & 295.1 & -95.9 & 486.6 & -201.8 \\
-4.3 & -579.9 & 152.8 & -615.3 & 64.3 & -595.3 & 84.2 & -599.9 & 95.9 & -771.5 & -102.8 & 1.2{\bullet}10^3 & -95.9 & 771.5 & -194.2 & 1.2{\bullet}10^3 \\
-246.1 & 8.5 & -507.8 & 292.3 & -360.3 & 132.4 & -393.6 & 168.4 & -486.6 & 194.2 & 677.7 & -218.2 & 486.6 & -194.2 & 830.0 & -383.4 \\
34.7 & -888.0 & 277.9 & -979.6 & 140.9 & -928.0 & 171.8 & -939.6 & 201.8 & -1.2{\bullet}10^3 & -241.8 & 1.8{\bullet}10^3 & -201.8 & 1.2{\bullet}10^3 & -383.4 & 1.9{\bullet}10^3
\end{bmatrix}$$

Gauss Point c

$$Ac := \frac{1}{.30635} \cdot \begin{bmatrix} .30635 & -.19365 & 0 & 0 \\ 0 & 0 & 0 & 1 \\ 0 & 1 & .30635 & -.19365 \end{bmatrix}$$

$$Gc := \begin{bmatrix} 0.3436 & 0.0 & 1.0309 & 0.0 & 0.0436 & 0.0 & 0.1309 & 0.0 & -1.3746 & 0.0 & 0.2 & 0.0 & -0.1746 & 0.0 & -0.2 & 0.0 \\ -0.0436 & 0.0 & -1.0309 & 0.0 & -0.3436 & 0.0 & -0.1309 & 0.0 & -0.2 & 0.0 & 1.3746 & 0.0 & 0.2 & 0.0 & 0.1746 & 0.0 \\ 0.0 & 0.3436 & 0.0 & 1.0309 & 0.0 & 0.0436 & 0.0 & 0.1309 & 0.0 & -1.3746 & 0.0 & 0.2 & 0.0 & -0.1746 & 0.0 & -0.2 \\ 0.0 & -0.0436 & 0.0 & -1.0309 & 0.0 & -0.3436 & 0.0 & -0.1309 & 0.0 & -0.2 & 0.0 & 1.3746 & 0.0 & 0.2 & 0.0 & 0.1746 \end{bmatrix}$$

$$Bc := Ac \cdot Gc$$

$$Kc := Bc^T \cdot D \cdot Bc \cdot .30635 \cdot .02$$

$$Kc = \begin{bmatrix}
67.6 & -16.0 & 369.6 & -213.9 & 71.2 & -64.3 & 46.9 & -27.2 & -201.0 & -4.9 & -220.1 & 248.7 & -67.3 & 40.9 & -67.0 & 36.8 \\
-16.0 & 31.9 & -237.5 & 325.4 & -73.2 & 90.3 & -30.2 & 41.3 & -14.7 & -32.3 & 285.3 & -338.5 & 45.6 & -61.6 & 40.7 & -56.7 \\
369.6 & -237.5 & 3.2 \cdot 10^3 & -1.7 \cdot 10^3 & 821.1 & -407.5 & 402.5 & -218.0 & -621.1 & 532.1 & -3.0 \cdot 10^3 & 1.4 \cdot 10^3 & -595.1 & 319.2 & -556.8 & 304.8 \\
-213.9 & 325.4 & -1.7 \cdot 10^3 & 5.7 \cdot 10^3 & -431.0 & 1.8 \cdot 10^3 & -218.0 & 729.6 & 408.5 & 682.1 & 1.5 \cdot 10^3 & -7.2 \cdot 10^3 & 321.2 & -1.1 \cdot 10^3 & 302.8 & -980.1 \\
71.2 & -73.2 & 821.1 & -431.0 & 237.2 & -88.7 & 104.3 & -54.7 & -32.3 & 204.8 & -903.2 & 286.3 & -156.2 & 79.0 & -142.2 & 77.6 \\
-64.3 & 90.3 & -407.5 & 1.8 \cdot 10^3 & -88.7 & 598.0 & -51.7 & 232.7 & 168.1 & 288.5 & 296.1 & -2.4 \cdot 10^3 & 75.1 & -354.5 & 73.0 & -311.5 \\
46.9 & -30.2 & 402.5 & -218.0 & 104.3 & -51.7 & 51.1 & -27.7 & -78.9 & 67.6 & -379.8 & 180.9 & -75.6 & 40.5 & -70.7 & 38.7 \\
-27.2 & 41.3 & -218.0 & 729.6 & -54.7 & 232.7 & -27.7 & 92.6 & 51.9 & 86.6 & 196.5 & -917.9 & 40.8 & -140.7 & 38.4 & -124.5 \\
-201.0 & -14.7 & -621.1 & 408.5 & -32.3 & 168.1 & -78.9 & 51.9 & 796.5 & 247.1 & -88.8 & -712.6 & 105.7 & -82.0 & 120.0 & -66.5 \\
-4.9 & -32.3 & 532.1 & 682.1 & 204.8 & 288.5 & 67.6 & 86.6 & 247.1 & 453.3 & -853.4 & -1.2 \cdot 10^3 & -105.6 & -137.5 & -87.8 & -110.3 \\
-220.1 & 285.3 & -3.0 \cdot 10^3 & 1.5 \cdot 10^3 & -903.2 & 296.1 & -379.8 & 196.5 & -88.8 & -853.4 & 3.5 \cdot 10^3 & -910.2 & 572.3 & -281.7 & 514.5 & -280.8 \\
248.7 & -338.5 & 1.4 \cdot 10^3 & -7.2 \cdot 10^3 & 286.3 & -2.4 \cdot 10^3 & 180.9 & -917.9 & -712.6 & -1.2 \cdot 10^3 & -910.2 & 9.5 \cdot 10^3 & -260.4 & 1.4 \cdot 10^3 & -257.2 & 1.2 \cdot 10^3 \\
-67.3 & 45.6 & -595.1 & 321.2 & -156.2 & 75.1 & -75.6 & 40.8 & 105.7 & -105.6 & 572.3 & -260.4 & 111.9 & -59.6 & 104.4 & -57.1 \\
40.9 & -61.6 & 319.2 & -1.1 \cdot 10^3 & 79.0 & -354.5 & 40.5 & -140.7 & -82.0 & -137.5 & -281.7 & 1.4 \cdot 10^3 & -59.6 & 213.7 & -56.4 & 188.9 \\
-67.0 & 40.7 & -556.8 & 302.8 & -142.2 & 73.0 & -70.7 & 38.4 & 120.0 & -87.8 & 514.5 & -257.2 & 104.4 & -56.4 & 98.0 & -53.6 \\
36.8 & -56.7 & 304.8 & -980.1 & 77.6 & -311.5 & 38.7 & -124.5 & -66.5 & -110.3 & -280.8 & 1.2 \cdot 10^3 & -57.1 & 188.9 & -53.6 & 167.3
\end{bmatrix}$$

GAUSS POINT d

$$Ad := \frac{1}{.69365} \cdot \begin{bmatrix} .69365 & 0 & 0 & 0 \\ 0 & 0 & 0 & 1 \\ 0 & 1 & .69365 & 0 \end{bmatrix}$$

$$Gd := \begin{bmatrix} -0.3873 & 0.0 & -0.3873 & 0.0 & -0.3873 & 0.0 & -0.3873 & 0.0 & 0.7746 & 0.0 & 0.5 & 0.0 & 0.7746 & 0.0 & -0.5 & 0.0 \\ -0.34365 & 0.0 & 0.04365 & 0.0 & -0.04365 & 0.0 & 0.34365 & 0.0 & -0.2 & 0.0 & 0.0 & 0.0 & 0.2 & 0.0 & 0.0 & 0.0 \\ 0.0 & -0.3873 & 0.0 & -0.3873 & 0.0 & -0.3873 & 0.0 & -0.3873 & 0.0 & 0.7746 & 0.0 & 0.5 & 0.0 & 0.7746 & 0.0 & -0.5 \\ 0.0 & -0.34365 & 0.0 & 0.04365 & 0.0 & -0.04365 & 0.0 & 0.34365 & 0.0 & -0.2 & 0.0 & 0.0 & 0.0 & 0.2 & 0.0 & 0.0 \end{bmatrix}$$

$$Bd := Ad \cdot Gd$$

$$Kd := Bd^T \cdot D \cdot Bd \cdot .69365 \cdot .02$$

$$Kd = \begin{bmatrix} 249.2 & 131.8 & 146.9 & 63.2 & 170.0 & 78.7 & 67.7 & 10.1 & -264.1 & -106.5 & -204.6 & -91.6 & -369.7 & -177.3 & 204.6 & 91.6 \\ 131.8 & 314.8 & 51.8 & 22.5 & 69.8 & 88.4 & -10.1 & -203.8 & -80.3 & 40.0 & -78.5 & -71.6 & -162.9 & -261.8 & 78.5 & 71.6 \\ 146.9 & 51.8 & 159.9 & -16.7 & 157.0 & -1.3 & 170.0 & -69.8 & -323.6 & 53.4 & -204.6 & 11.6 & -310.2 & -17.4 & 204.6 & -11.6 \\ 63.2 & 22.5 & -16.7 & 59.6 & 1.3 & 51.3 & -78.7 & 88.4 & 56.7 & -130.1 & 10 & -71.6 & -25.8 & -91.8 & -10 & 71.6 \\ 170.0 & 69.8 & 157.0 & 1.3 & 159.9 & 16.7 & 146.9 & -51.8 & -310.2 & 17.4 & -204.6 & -11.6 & -323.6 & -53.4 & 204.6 & 11.6 \\ 78.7 & 88.4 & -1.3 & 51.3 & 16.7 & 59.6 & -63.2 & 22.5 & 25.8 & -91.8 & -10 & -71.6 & -56.7 & -130.1 & 10 & 71.6 \\ 67.7 & -10.1 & 170.0 & -78.7 & 146.9 & -63.2 & 249.2 & -131.8 & -369.7 & 177.3 & -204.6 & 91.6 & -264.1 & 106.5 & 204.6 & -91.6 \\ 10.1 & -203.8 & -69.8 & 88.4 & -51.8 & 22.5 & -131.8 & 314.8 & 162.9 & -261.8 & 78.5 & -71.6 & 80.3 & 40.0 & -78.5 & 71.6 \\ -264.1 & -80.3 & -323.6 & 56.7 & -310.2 & 25.8 & -369.7 & 162.9 & 664.6 & -153.4 & 409.2 & -53.3 & 603.1 & -11.8 & -409.2 & 53.3 \\ -106.5 & 40.0 & 53.4 & -130.1 & 17.4 & -91.8 & 177.3 & -261.8 & -153.4 & 309.7 & -45.7 & 143.2 & 11.8 & 134.0 & 45.7 & -143.2 \\ -204.6 & -78.5 & -204.6 & 10 & -204.6 & -10 & -204.6 & 78.5 & 409.2 & -45.7 & 264.1 & 0.0 & 409.2 & 45.7 & -264.1 & 0.0 \\ -91.6 & -71.6 & 11.6 & -71.6 & -11.6 & -71.6 & 91.6 & -71.6 & -53.3 & 143.2 & 0.0 & 92.4 & 53.3 & 143.2 & 0.0 & -92.4 \\ -369.7 & -162.9 & -310.2 & -25.8 & -323.6 & -56.7 & -264.1 & 80.3 & 603.1 & 11.8 & 409.2 & 53.3 & 664.6 & 153.4 & -409.2 & -53.3 \\ -177.3 & -261.8 & -17.4 & -91.8 & -53.4 & -130.1 & 106.5 & 40.0 & -11.8 & 134.0 & 45.7 & 143.2 & 153.4 & 309.7 & -45.7 & -143.2 \\ 204.6 & 78.5 & 204.6 & -10 & 204.6 & 10 & 204.6 & -78.5 & -409.2 & 45.7 & -264.1 & 0.0 & -409.2 & -45.7 & 264.1 & 0.0 \\ 91.6 & 71.6 & -11.6 & 71.6 & 11.6 & 71.6 & -91.6 & 71.6 & 53.3 & -143.2 & 0.0 & -92.4 & -53.3 & -143.2 & 0.0 & 92.4 \end{bmatrix}$$

GAUSS POINT e

$$Ae := \frac{1}{0.5} \cdot \begin{bmatrix} .5 & 0 & 0 & 0 \\ 0 & 0 & 0 & 1 \\ 0 & 1 & .5 & 0 \end{bmatrix}$$

$$Ge := \begin{bmatrix} 0.0 & 0.0 & 0.0 & 0.0 & 0.0 & 0.0 & 0.0 & 0.0 & 0.0 & 0.0 & 0.5 & 0.0 & 0.0 & 0.0 & -0.5 & 0.0 \\ 0.0 & 0.0 & 0.0 & 0.0 & 0.0 & 0.0 & 0.0 & 0.0 & -0.5 & 0.0 & 0.0 & 0.0 & 0.5 & 0.0 & 0.0 & 0.0 \\ 0.0 & 0.0 & 0.0 & 0.0 & 0.0 & 0.0 & 0.0 & 0.0 & 0.0 & 0.0 & 0.0 & 0.5 & 0.0 & 0.0 & 0.0 & -0.5 \\ 0.0 & 0.0 & 0.0 & 0.0 & 0.0 & 0.0 & 0.0 & 0.0 & 0.0 & -0.5 & 0.0 & 0.0 & 0.0 & 0.5 & 0.0 & 0.0 \end{bmatrix}$$

$$Be := Ae \cdot Ge$$

$$Ke := Be^T \cdot D \cdot Be \cdot 0.5 \cdot .02$$

$$Ke = \begin{bmatrix} 0.0 & 0.0 & 0.0 & 0.0 & 0.0 & 0.0 & 0.0 & 0.0 & 0.0 & 0.0 & 0.0 & 0.0 & 0.0 & 0.0 & 0.0 & 0.0 \\ 0.0 & 0.0 & 0.0 & 0.0 & 0.0 & 0.0 & 0.0 & 0.0 & 0.0 & 0.0 & 0.0 & 0.0 & 0.0 & 0.0 & 0.0 & 0.0 \\ 0.0 & 0.0 & 0.0 & 0.0 & 0.0 & 0.0 & 0.0 & 0.0 & 0.0 & 0.0 & 0.0 & 0.0 & 0.0 & 0.0 & 0.0 & 0.0 \\ 0.0 & 0.0 & 0.0 & 0.0 & 0.0 & 0.0 & 0.0 & 0.0 & 0.0 & 0.0 & 0.0 & 0.0 & 0.0 & 0.0 & 0.0 & 0.0 \\ 0.0 & 0.0 & 0.0 & 0.0 & 0.0 & 0.0 & 0.0 & 0.0 & 0.0 & 0.0 & 0.0 & 0.0 & 0.0 & 0.0 & 0.0 & 0.0 \\ 0.0 & 0.0 & 0.0 & 0.0 & 0.0 & 0.0 & 0.0 & 0.0 & 0.0 & 0.0 & 0.0 & 0.0 & 0.0 & 0.0 & 0.0 & 0.0 \\ 0.0 & 0.0 & 0.0 & 0.0 & 0.0 & 0.0 & 0.0 & 0.0 & 0.0 & 0.0 & 0.0 & 0.0 & 0.0 & 0.0 & 0.0 & 0.0 \\ 0.0 & 0.0 & 0.0 & 0.0 & 0.0 & 0.0 & 0.0 & 0.0 & 0.0 & 0.0 & 0.0 & 0.0 & 0.0 & 0.0 & 0.0 & 0.0 \\ 0.0 & 0.0 & 0.0 & 0.0 & 0.0 & 0.0 & 0.0 & 0.0 & 266.5 & 0.0 & 0.0 & -133.3 & -266.5 & 0.0 & 0.0 & 133.3 \\ 0.0 & 0.0 & 0.0 & 0.0 & 0.0 & 0.0 & 0.0 & 0.0 & 0.0 & 761.5 & -114.2 & 0.0 & 0.0 & -761.5 & 114.2 & 0.0 \\ 0.0 & 0.0 & 0.0 & 0.0 & 0.0 & 0.0 & 0.0 & 0.0 & 0.0 & -114.2 & 190.4 & 0.0 & 0.0 & 114.2 & -190.4 & 0.0 \\ 0.0 & 0.0 & 0.0 & 0.0 & 0.0 & 0.0 & 0.0 & 0.0 & -133.3 & 0.0 & 0.0 & 66.6 & 133.3 & 0.0 & 0.0 & -66.6 \\ 0.0 & 0.0 & 0.0 & 0.0 & 0.0 & 0.0 & 0.0 & 0.0 & -266.5 & 0.0 & 0.0 & 133.3 & 266.5 & 0.0 & 0.0 & -133.3 \\ 0.0 & 0.0 & 0.0 & 0.0 & 0.0 & 0.0 & 0.0 & 0.0 & 0.0 & -761.5 & 114.2 & 0.0 & 0.0 & 761.5 & -114.2 & 0.0 \\ 0.0 & 0.0 & 0.0 & 0.0 & 0.0 & 0.0 & 0.0 & 0.0 & 0.0 & 114.2 & -190.4 & 0.0 & 0.0 & -114.2 & 190.4 & 0.0 \\ 0.0 & 0.0 & 0.0 & 0.0 & 0.0 & 0.0 & 0.0 & 0.0 & 133.3 & 0.0 & 0.0 & -66.6 & -133.3 & 0.0 & 0.0 & 66.6 \end{bmatrix}$$

GAUSS POINT f

$$Af := \frac{1}{.30635} \cdot \begin{bmatrix} .30635 & 0 & 0 & 0 \\ 0 & 0 & 0 & 1 \\ 0 & 1 & .30635 & 0 \end{bmatrix}$$

$$Gf := \begin{bmatrix} 0.3873 & 0.0 & 0.3873 & 0.0 & 0.3873 & 0.0 & 0.3873 & 0.0 & -0.7746 & 0.0 & 0.5 & 0.0 & -0.7746 & 0.0 & -0.5 & 0.0 \\ 0.04365 & 0.0 & -0.34365 & 0.0 & 0.34365 & 0.0 & -0.04365 & 0.0 & -0.2 & 0.0 & 0.0 & 0.0 & 0.2 & 0.0 & 0.0 & 0.0 \\ 0.0 & 0.3873 & 0.0 & 0.3873 & 0.0 & 0.3873 & 0.0 & 0.3873 & 0.0 & -0.7746 & 0.0 & 0.5 & 0.0 & -0.7746 & 0.0 & -0.5 \\ 0.0 & 0.04365 & 0.0 & -0.34365 & 0.0 & 0.34365 & 0.0 & -0.04365 & 0.0 & -0.2 & 0.0 & 0.0 & 0.0 & 0.2 & 0.0 & 0.0 \end{bmatrix}$$

$$Bf := Af \cdot Gf$$

$$Kf := Bf^T \cdot D \cdot Bf \cdot .30635 \cdot .02$$

$$Kf = \begin{bmatrix} 73.3 & 16.7 & 43.9 & -51.8 & 96.1 & 69.8 & 66.7 & 1.3 & -155.2 & -53.4 & 90.4 & 11.6 & -124.8 & 17.4 & -90.4 & -11.6 \\ 16.7 & 34.0 & -63.2 & -50.1 & 78.7 & 99.1 & -1.3 & 15.0 & -56.7 & -92.4 & 10 & 31.6 & 25.8 & -5.6 & -10 & -31.6 \\ 43.9 & -63.2 & 275.5 & -131.8 & -135.5 & -10.1 & 96.1 & -78.7 & -20.4 & 106.5 & 90.4 & -91.6 & -259.6 & 177.3 & -90.4 & 91.6 \\ -51.8 & -50.1 & -131.8 & 611.6 & 10.1 & -562.6 & -69.8 & 99.1 & 80.3 & 292.7 & -78.5 & 31.6 & 162.9 & -390.7 & 78.5 & -31.6 \\ 96.1 & 78.7 & -135.5 & 10.1 & 275.5 & 131.8 & 43.9 & 63.2 & -259.6 & -177.3 & 90.4 & 91.6 & -20.4 & -106.5 & -90.4 & -91.6 \\ 69.8 & 99.1 & -10.1 & -562.6 & 131.8 & 611.6 & 51.8 & -50.1 & -162.9 & -390.7 & 78.5 & 31.6 & -80.3 & 292.7 & -78.5 & -31.6 \\ 66.7 & -1.3 & 96.1 & -69.8 & 43.9 & 51.8 & 73.3 & -16.7 & -124.8 & -17.4 & 90.4 & -11.6 & -155.2 & 53.4 & -90.4 & 11.6 \\ 1.3 & 15.0 & -78.7 & 99.1 & 63.2 & -50.1 & -16.7 & 34.0 & -25.8 & -5.6 & -10 & 31.6 & 56.7 & -92.4 & 10 & -31.6 \\ -155.2 & -56.7 & -20.4 & 80.3 & -259.6 & -162.9 & -124.8 & -25.8 & 349.5 & 153.4 & -180.7 & -53.3 & 210.3 & 11.8 & 180.7 & 53.3 \\ -53.4 & -92.4 & 106.5 & 292.7 & -177.3 & -390.7 & -17.4 & -5.6 & 153.4 & 296.8 & -45.7 & -63.2 & -11.8 & -100.9 & 45.7 & 63.2 \\ 90.4 & 10 & 90.4 & -78.5 & 90.4 & 78.5 & 90.4 & -10 & -180.7 & -45.7 & 116.6 & 0.0 & -180.7 & 45.7 & -116.6 & 0.0 \\ 11.6 & 31.6 & -91.6 & 31.6 & 91.6 & 31.6 & -11.6 & 31.6 & -53.3 & -63.2 & 0.0 & 40.8 & 53.3 & -63.2 & 0.0 & -40.8 \\ -124.8 & 25.8 & -259.6 & 162.9 & -20.4 & -80.3 & -155.2 & 56.7 & 210.3 & -11.8 & -180.7 & 53.3 & 349.5 & -153.4 & 180.7 & -53.3 \\ 17.4 & -5.6 & 177.3 & -390.7 & -106.5 & 292.7 & 53.4 & -92.4 & 11.8 & -100.9 & 45.7 & -63.2 & -153.4 & 296.8 & -45.7 & 63.2 \\ -90.4 & -10 & -90.4 & 78.5 & -90.4 & -78.5 & -90.4 & 10 & 180.7 & 45.7 & -116.6 & 0.0 & 180.7 & -45.7 & 116.6 & 0.0 \\ -11.6 & -31.6 & 91.6 & -31.6 & -91.6 & -31.6 & 11.6 & -31.6 & 53.3 & 63.2 & 0.0 & -40.8 & -53.3 & 63.2 & 0.0 & 40.8 \end{bmatrix}$$

GAUSS POINT g

$$Ag := \frac{1}{.69365} \cdot \begin{bmatrix} .69365 & .19365 & 0 & 0 \\ 0 & 0 & 0 & 1 \\ 0 & 1 & .69365 & .19365 \end{bmatrix}$$

$$Gg := \begin{bmatrix} -0.0436 & 0.0 & -0.1309 & 0.0 & -0.3436 & 0.0 & -1.0309 & 0.0 & 0.1746 & 0.0 & 0.2 & 0.0 & 1.3746 & 0.0 & -0.2 & 0.0 \\ 0.34365 & 0.0 & 0.13095 & 0.0 & 0.04365 & 0.0 & 1.03095 & 0.0 & -0.2 & 0.0 & -0.1746 & 0.0 & 0.2 & 0.0 & -1.3746 & 0.0 \\ 0.0 & -0.0436 & 0.0 & -0.1309 & 0.0 & -0.3436 & 0.0 & -1.0309 & 0.0 & 0.1746 & 0.0 & 0.2 & 0.0 & 1.3746 & 0.0 & -0.2 \\ 0.0 & 0.34365 & 0.0 & 0.13095 & 0.0 & 0.04365 & 0.0 & 1.03095 & 0.0 & -0.2 & 0.0 & -0.1746 & 0.0 & 0.2 & 0.0 & -1.3746 \end{bmatrix}$$

$$Bg := Ag \cdot Gg$$

$$Kg := Bg^T \cdot D \cdot Bg \cdot .69365 \cdot .02$$

$$Kg = \begin{bmatrix}
93.6 & 17.8 & 29.4 & -14.2 & -6.8 & -59.7 & 231.2 & -111.5 & -46.2 & 17.0 & -37.7 & 23.5 & 131.9 & 266.8 & -395.3 & -139.8 \\
17.8 & 260.3 & -11.2 & 97.0 & -50.8 & 26.5 & -87.9 & 763.5 & 13.1 & -148.6 & 18.9 & -128.8 & 230.2 & 178.6 & -130.0 & -1.0 \bullet 10^3 \\
29.4 & -11.2 & 22.6 & -12.2 & 37.4 & -25.0 & 177.8 & -96.3 & -32.0 & 16.9 & -32.6 & 18.1 & -122.4 & 91.2 & -80.1 & 18.5 \\
-14.2 & 97.0 & -12.2 & 40.9 & -22.0 & 24.1 & -96.3 & 322.3 & 17.2 & -61.6 & 17.8 & -55.5 & 75.5 & 7.6 & 34.2 & -374.9 \\
-6.8 & -50.8 & 37.4 & -22.0 & 117.5 & -14.3 & 294.7 & -173.4 & -48.3 & 33.0 & -58.8 & 30.0 & -494.1 & 3.0 & 158.3 & 194.6 \\
-59.7 & 26.5 & -25.0 & 24.1 & -14.3 & 44.8 & -196.9 & 189.9 & 37.7 & -33.7 & 33.9 & -35.3 & -6.8 & -156.1 & 231.2 & -60.2 \\
231.2 & -87.9 & 177.8 & -96.3 & 294.7 & -196.9 & 1.4 \bullet 10^3 & -758.4 & -251.7 & 133.2 & -257.1 & 142.4 & -964.5 & 718.2 & -630.8 & 145.9 \\
-111.5 & 763.5 & -96.3 & 322.3 & -173.4 & 189.9 & -758.4 & 2.5 \bullet 10^3 & 135.2 & -485.3 & 140.4 & -436.8 & 594.6 & 59.7 & 269.5 & -3.0 \bullet 10^3 \\
-46.2 & 13.1 & -32.0 & 17.2 & -48.3 & 37.7 & -251.7 & 135.2 & 45.6 & -23.5 & 45.8 & -25.6 & 148.7 & -141.6 & 138.0 & -12.4 \\
17.0 & -148.6 & 16.9 & -61.6 & 33.0 & -33.7 & 133.2 & -485.3 & -23.5 & 93.0 & -24.9 & 83.3 & -118.1 & -25.0 & -33.7 & 578.0 \\
-37.7 & 18.9 & -32.6 & 17.8 & -58.8 & 33.9 & -257.1 & 140.4 & 45.8 & -24.9 & 47.6 & -26.1 & 201.7 & -119.3 & 91.2 & -40.7 \\
23.5 & -128.8 & 18.1 & -55.5 & 30.0 & -35.3 & 142.4 & -436.8 & -25.6 & 83.3 & -26.1 & 75.4 & -98.0 & 3.3 & -64.3 & 494.3 \\
131.9 & 230.2 & -122.4 & 75.5 & -494.1 & -6.8 & -964.5 & 594.6 & 148.7 & -118.1 & 201.7 & -98.0 & 2.2 \bullet 10^3 & 283.2 & -1.1 \bullet 10^3 & -960.6 \\
266.8 & 178.6 & 91.2 & 7.6 & 3.0 & -156.1 & 718.2 & 59.7 & -141.6 & -25.0 & -119.3 & 3.3 & 283.2 & 844.4 & -1.1 \bullet 10^3 & -912.4 \\
-395.3 & -130.0 & -80.1 & 34.2 & 158.3 & 231.2 & -630.8 & 269.5 & 138.0 & -33.7 & 91.2 & -64.3 & -1.1 \bullet 10^3 & -1.1 \bullet 10^3 & 1.8 \bullet 10^3 & 794.4 \\
-139.8 & -1.0 \bullet 10^3 & 18.5 & -374.9 & 194.6 & -60.2 & 145.9 & -3.0 \bullet 10^3 & -12.4 & 578.0 & -40.7 & 494.3 & -960.6 & -912.4 & 794.4 & 4.3 \bullet 10^3
\end{bmatrix}$$

165

BASIC PRINCIPLES OF THE FINITE ELEMENT METHOD

GAUSS POINT h

$$Ah := \frac{1}{.5} \cdot \begin{bmatrix} .5 & .19365 & 0 & 0 \\ 0 & 0 & 0 & 1 \\ 0 & 1 & .5 & .19365 \end{bmatrix}$$

$$Gh := \begin{bmatrix} 0.0436 & 0.0 & -0.0436 & 0.0 & 0.3436 & 0.0 & -0.3436 & 0.0 & 0.0 & 0.0 & 0.2 & 0.0 & 0.0 & 0.0 & -0.2 & 0.0 \\ 0.3873 & 0.0 & 0.3873 & 0.0 & 0.3873 & 0.0 & 0.3873 & 0.0 & -0.5 & 0.0 & -0.7746 & 0.0 & 0.5 & 0.0 & -0.7746 & 0.0 \\ 0.0 & 0.0436 & 0.0 & -0.0436 & 0.0 & 0.3436 & 0.0 & -0.3436 & 0.0 & 0.0 & 0.0 & 0.2 & 0.0 & 0.0 & 0.0 & -0.2 \\ 0.0 & 0.3873 & 0.0 & 0.3873 & 0.0 & 0.3873 & 0.0 & 0.3873 & 0.0 & -0.5 & 0.0 & -0.7746 & 0.0 & 0.5 & 0.0 & -0.7746 \end{bmatrix}$$

$$Bh := Ah \cdot Gh$$

$$Kh := Bh^T \cdot D \cdot Bh \cdot 0.5 \cdot .02$$

$$Kh = \begin{bmatrix} 188.5 & 74.2 & 175.6 & 56.2 & 232.7 & 136.2 & 131.4 & -5.7 & -235.0 & -84.2 & -334.6 & -89.2 & 235.0 & 84.2 & -393.5 & -171.7 \\ 74.2 & 466.9 & 58.8 & 462.4 & 127.3 & 482.4 & 5.7 & 446.9 & -85.9 & -599.8 & -97.6 & -919.0 & 85.9 & 599.8 & -168.4 & -939.6 \\ 175.6 & 58.8 & 168.5 & 40.8 & 199.9 & 120.7 & 144.2 & -21.1 & -222.1 & -64.3 & -327.9 & -58.3 & 222.1 & 64.3 & -360.3 & -140.9 \\ 56.2 & 462.4 & 40.8 & 459.9 & 109.3 & 470.9 & -12.3 & 451.4 & -62.6 & -595.3 & -61.6 & -916.6 & 62.6 & 595.3 & -132.4 & -928.0 \\ 232.7 & 127.3 & 199.9 & 109.3 & 345.4 & 189.2 & 87.1 & 47.4 & -279.2 & -152.7 & -357.4 & -195.3 & 279.2 & 152.7 & -507.8 & -277.9 \\ 136.2 & 482.4 & 120.7 & 470.9 & 189.2 & 521.8 & 67.6 & 431.4 & -165.8 & -615.3 & -221.5 & -927.0 & 165.8 & 615.3 & -292.3 & -979.6 \\ 131.4 & 5.7 & 144.2 & -12.3 & 87.1 & 67.6 & 188.5 & -74.2 & -177.9 & 4.2 & -305.1 & 47.9 & 177.9 & -4.2 & -246.1 & -34.7 \\ -5.7 & 446.9 & -21.1 & 451.4 & 47.4 & 431.4 & -74.2 & 466.9 & 17.3 & -579.9 & 62.2 & -908.6 & -17.3 & 579.9 & -8.5 & -888.0 \\ -235.0 & -85.9 & -222.1 & -62.6 & -279.2 & -165.8 & -177.9 & 17.3 & 295.1 & 95.9 & 427.6 & 95.2 & -295.1 & -95.9 & 486.6 & 201.8 \\ -84.2 & -599.8 & -64.3 & -595.3 & -152.7 & -615.3 & 4.2 & -579.9 & 95.9 & 771.5 & 102.8 & 1.2 \bullet 10^3 & -95.9 & -771.5 & 194.2 & 1.2 \bullet 10^3 \\ -334.6 & -97.6 & -327.9 & -61.6 & -357.4 & -221.5 & -305.1 & 62.2 & 427.6 & 102.8 & 647.3 & 76.7 & -427.6 & -102.8 & 677.7 & 241.8 \\ -89.2 & -919.0 & -58.3 & -916.6 & -195.3 & -927.0 & 47.9 & -908.6 & 95.2 & 1.2 \bullet 10^3 & 76.7 & 1.8 \bullet 10^3 & -95.2 & -1.2 \bullet 10^3 & 218.2 & 1.8 \bullet 10^3 \\ 235.0 & 85.9 & 222.1 & 62.6 & 279.2 & 165.8 & 177.9 & -17.3 & -295.1 & -95.9 & -427.6 & -95.2 & 295.1 & 95.9 & -486.6 & -201.8 \\ 84.2 & 599.8 & 64.3 & 595.3 & 152.7 & 615.3 & -4.2 & 579.9 & -95.9 & -771.5 & -102.8 & -1.2 \bullet 10^3 & 95.9 & 771.5 & -194.2 & -1.2 \bullet 10^3 \\ -393.5 & -168.4 & -360.3 & -132.4 & -507.8 & -292.3 & -246.1 & -8.5 & 486.6 & 194.2 & 677.7 & 218.2 & -486.6 & -194.2 & 830.0 & 383.4 \\ -171.7 & -939.6 & -140.9 & -928.0 & -277.9 & -979.6 & -34.7 & -888.0 & 201.8 & 1.2 \bullet 10^3 & 241.8 & 1.8 \bullet 10^3 & -201.8 & -1.2 \bullet 10^3 & 383.4 & 1.9 \bullet 10^3 \end{bmatrix}$$

GAUSS POINT k

$$Ak := \frac{1}{.30635} \cdot \begin{bmatrix} .30635 & .19365 & 0 & 0 \\ 0 & 0 & 0 & 1 \\ 0 & 1 & .30635 & .19365 \end{bmatrix}$$

$$Gk := \begin{bmatrix} 0.1309 & 0.0 & 0.0436 & 0.0 & 1.0309 & 0.0 & 0.3436 & 0.0 & -0.1746 & 0.0 & 0.2 & 0.0 & -1.3746 & 0.0 & -0.2 & 0.0 \\ 0.1309 & 0.0 & 0.3436 & 0.0 & 1.0309 & 0.0 & 0.0436 & 0.0 & -0.2 & 0.0 & -1.3746 & 0.0 & 0.2 & 0.0 & -0.1746 & 0.0 \\ 0.0 & 0.1309 & 0.0 & 0.0436 & 0.0 & 1.0309 & 0.0 & 0.3436 & 0.0 & -0.1746 & 0.0 & 0.2 & 0.0 & -1.3746 & 0.0 & -0.2 \\ 0.0 & 0.1309 & 0.0 & 0.3436 & 0.0 & 1.0309 & 0.0 & 0.0436 & 0.0 & -0.2 & 0.0 & -1.3746 & 0.0 & 0.2 & 0.0 & -0.1746 \end{bmatrix}$$

$$Bk := Ak \cdot Gk$$

$$Kk := Bk^T \cdot D \cdot Bk \cdot .30635 \cdot .02$$

166

$$
Kk = \begin{bmatrix}
51.1 & 27.7 & 104.3 & 51.7 & 402.5 & 218.0 & 46.9 & 30.2 & -75.6 & -40.5 & -379.8 & -180.9 & -78.9 & -67.6 & -70.7 & -38.7 \\
27.7 & 92.6 & 54.7 & 232.7 & 218.0 & 729.6 & 27.2 & 41.3 & -40.8 & -140.7 & -196.5 & -917.9 & -51.9 & 86.6 & -38.4 & -124.5 \\
104.3 & 54.7 & 237.2 & 88.7 & 821.1 & 431.0 & 71.2 & 73.2 & -156.2 & -79.0 & -903.2 & -286.3 & -32.3 & -204.8 & -142.2 & -77.6 \\
51.7 & 232.7 & 88.7 & 598.0 & 407.5 & 1.8 \bullet 10^3 & 64.3 & 90.3 & -75.1 & -354.5 & -296.1 & -2.4 \bullet 10^3 & -168.1 & 288.5 & -73.0 & -311.5 \\
402.5 & 218.0 & 821.1 & 407.5 & 3.2 \bullet 10^3 & 1.7 \bullet 10^3 & 369.6 & 237.5 & -595.1 & -319.2 & -3.0 \bullet 10^3 & -1.4 \bullet 10^3 & -621.1 & -532.1 & -556.8 & -304.8 \\
218.0 & 729.6 & 431.0 & 1.8 \bullet 10^3 & 1.7 \bullet 10^3 & 5.7 \bullet 10^3 & 213.9 & 325.4 & -321.2 & -1.1 \bullet 10^3 & -1.5 \bullet 10^3 & -7.2 \bullet 10^3 & -408.5 & 682.1 & -302.8 & -980.1 \\
46.9 & 27.2 & 71.2 & 64.3 & 369.6 & 213.9 & 67.6 & 16.0 & -67.3 & -40.9 & -220.1 & -248.7 & -201.0 & 4.9 & -67.0 & -36.8 \\
30.2 & 41.3 & 73.2 & 90.3 & 237.5 & 325.4 & 16.0 & 31.9 & -45.6 & -61.6 & -285.3 & -338.5 & 14.7 & -32.3 & -40.7 & -56.7 \\
-75.6 & -40.8 & -156.2 & -75.1 & -595.1 & -321.2 & -67.3 & -45.6 & 111.9 & 59.6 & 572.3 & 260.4 & 105.7 & 105.6 & 104.4 & 57.1 \\
-40.5 & -140.7 & -79.0 & -354.5 & -319.2 & -1.1 \bullet 10^3 & -40.9 & -61.6 & 59.6 & 213.7 & 281.7 & 1.4 \bullet 10^3 & 82.0 & -137.5 & 56.4 & 188.9 \\
-379.8 & -196.5 & -903.2 & -296.1 & -3.0 \bullet 10^3 & -1.5 \bullet 10^3 & -220.1 & -285.3 & 572.3 & 281.7 & 3.5 \bullet 10^3 & 910.2 & -88.8 & 853.4 & 514.5 & 280.8 \\
-180.9 & -917.9 & -286.3 & -2.4 \bullet 10^3 & -1.4 \bullet 10^3 & -7.2 \bullet 10^3 & -248.7 & -338.5 & 260.4 & 1.4 \bullet 10^3 & 910.2 & 9.5 \bullet 10^3 & 712.6 & -1.2 \bullet 10^3 & 257.2 & 1.2 \bullet 10^3 \\
-78.9 & -51.9 & -32.3 & -168.1 & -621.1 & -408.5 & -201.0 & 14.7 & 105.7 & 82.0 & -88.8 & 712.6 & 796.5 & -247.1 & 120.0 & 66.5 \\
-67.6 & 86.6 & -204.8 & 288.5 & -532.1 & 682.1 & 4.9 & -32.3 & 105.6 & -137.5 & 853.4 & -1.2 \bullet 10^3 & -247.1 & 453.3 & 87.8 & -110.3 \\
-70.7 & -38.4 & -142.2 & -73.0 & -556.8 & -302.8 & -67.0 & -40.7 & 104.4 & 56.4 & 514.5 & 257.2 & 120.0 & 87.8 & 98.0 & 53.6 \\
-38.7 & -124.5 & -77.6 & -311.5 & -304.8 & -980.1 & -36.8 & -56.7 & 57.1 & 188.9 & 280.8 & 1.2 \bullet 10^3 & 66.5 & -110.3 & 53.6 & 167.3
\end{bmatrix}
$$

8.2.3 GAUSSIAN QUADRATURE INTEGRATION TO FORM THE ELEMENT STIFFNESS MATRIX

We now perform the integration of the function which we shall call ϕ

$$\phi = B^T \cdot D \cdot B \cdot \det J \cdot t$$

over the area of the element.

We have explained that this integration must be performed numerically and we use Gauss quadrature for this purpose. The integral is then

$$\phi_a \cdot w_a + \phi_b \cdot w_b + \phi_c \cdot w_c + \phi_d \cdot w_d + \phi_e \cdot w_e + \phi_f \cdot w_f + \phi_g \cdot w_g + \phi_h \cdot w_h + \phi_k \cdot w_k$$

where for example ϕ_d is the value at gauss point d of the function to be integrated and w_d is the weight attached to that Gauss point.

We have determined the weights (see Fig. 7.24 in Section 7.2.8). They are

$$w_a = w_c = w_g = w_k = 25/81$$

$$w_b = w_d = w_f = w_h = 40/81$$

$$w_e = 64/81$$

The values of ϕ are the K matrices derived above for each Gauss point. For example $\phi_a = Ka$. so the integral of ϕ over the element is

$$64/81[Ke] + 40/81[Kb + Kd + Kf + Kh] + 25/81[Ka + Kc + Kg + Kk]$$

We call this sum KR in what follows.

KR is the sum of nine $[16 \times 16]$ matrices. To add the matrices we add all the nine terms in similar positions in each matrix, multiplying each term by the weight appropriate to the particular Gauss point.

We now proceed with the calculations.

$$KR := \frac{64 \cdot Ke + 40 \cdot (Kb + Kd + Kf + Kh) + 25 \cdot (Ka + Kc + Kg + Kk)}{81}$$

$$KR =$$

843.1	389.8	470.2	6.4	516.5	205.4	367.8	-6.3	-632.6	-348.9	-648.4	-122.9	-298.1	-6.5	-618.6	-117.0
389.8	1535.3	-6.3	688.5	205.4	914.4	6.3	844.9	-298.1	-116.6	-122.9	-1484.6	-6.5	-209.1	-167.7	-2173.1
470.2	-6.3	1563.7	-648.6	738.0	6.3	516.5	-205.4	-544.0	116.0	-1624.9	375.8	-555.9	239.4	-563.9	122.9
6.4	688.5	-648.6	2800.7	-6.3	1358.7	-205.4	914.4	65.2	124.1	426.5	-3922.8	239.4	-509.0	122.9	-1455.1
516.5	205.4	738.0	-6.3	1563.7	648.6	470.3	6.3	-555.9	-239.4	-1624.9	-375.8	-544.0	-116.0	-563.9	-122.9
205.4	914.4	6.3	1358.7	648.6	2800.7	-6.4	688.6	-239.4	-509.0	-426.5	-3922.7	-65.2	124.1	-122.9	-1455.1
367.8	6.3	516.5	-205.4	470.3	-6.4	843.1	-389.8	-298.1	6.5	-648.4	122.9	-632.5	348.9	-618.6	116.9
-6.3	844.9	-205.4	914.4	6.3	688.6	-389.8	1535.4	6.5	-209.2	122.9	-1484.6	298.1	-116.6	167.7	-2173.1
-632.6	-298.1	-544.0	65.2	-555.9	-239.4	-298.1	6.5	1973.9	0.0	338.4	-181.2	56.8	0.0	-338.4	647.0
-348.9	-116.6	116.0	124.1	-239.4	-509.0	6.5	-209.2	0.0	2158.3	-181.2	118.5	0.0	-1447.6	647.0	-118.5
-648.4	-122.9	-1624.9	426.5	-1624.9	-426.5	-648.4	122.9	338.4	-181.2	3165.4	0.0	338.4	181.2	704.8	0.0
-122.9	-1484.6	375.8	-3922.8	-375.8	-3922.7	122.9	-1484.6	-181.2	118.5	0.0	7816.3	181.2	118.5	0.0	2762.3
-298.1	-6.5	-555.9	239.4	-544.0	-65.2	-632.5	298.1	56.8	0.0	338.4	181.2	1973.9	0.0	-338.4	-647.0
-6.5	-209.1	239.4	-509.0	-116.0	124.1	348.9	-116.6	0.0	-1447.6	181.2	118.5	0.0	2158.3	-647.0	-118.5
-618.6	-167.7	-563.9	122.9	-563.9	-122.9	-618.6	167.7	-338.4	647.0	704.8	0.0	-338.4	-647.0	2337.2	0.0
-117.0	-2173.1	122.9	-1455.1	-122.9	-1455.1	116.9	-2173.1	647.0	-118.5	0.0	2762.3	-647.0	-118.5	0.0	4731.3

We have reduced the displayed precision of this large matrix in order to fit it on the page width.

To give an example to illustrate the process of summing the Gauss point matrices, look at the first term in the top left-hand corner of *KR*. It is 843.1. This is the sum of the nine top left-hand corner terms of the Gauss point matrices *Ka* to *Kk* multiplied by the Gauss point weight. If we extract the term from each matrix in turn we find that the sum is

$$64/81\{0\} + 40/81[188.5 + 249.2 + 73.3 + 188.5] + 25/81[1400.2 + 67.6 + 93.6 + 51.1] = 843.1$$

which agrees with the term 843.1 in *KR*.

8.2.4 THE CONDENSED STIFFNESS MATRIX AND ITS INVERSION TO GET THE NODAL DISPLACEMENTS

We now condense the matrix *KR* in exactly the same way that we did in the previous example by deleting rows 1, 2, 7, 8, 15 and 16 and columns 1, 2, 7, 8, 15 and 16. This gives the following [10 × 10] matrix which we call *KC*.

$$KC :=$$

1563.808	-648.642	738.106	6.345	-544.009	115.965	-1625.03	375.779	-555.945	239.398
-648.642	2800.944	-6.345	1358.938	65.205	124.058	426.539	-3922.995	239.398	-509.035
738.106	-6.345	1563.808	648.642	-555.945	-239.398	-1625.03	-375.779	-544.009	-115.965
6.345	1358.938	648.642	2800.944	-239.398	-509.035	-426.539	-3922.995	-65.205	124.058
-544.009	65.205	-555.945	-239.398	1973.937	0.0	338.45	-181.169	56.747	0.0
115.965	124.058	-239.398	-509.035	0.0	2158.328	-181.171	118.455	0.0	-1447.602
-1625.03	426.539	-1625.03	-426.539	338.45	-181.171	3165.456	0.0	338.45	181.171
375.779	-3922.995	-375.779	-3922.995	-181.169	118.455	0.0	7816.414	181.169	118.455
-555.945	239.398	-544.009	-65.205	56.747	0.0	338.45	181.169	1973.937	0.0
239.398	-509.035	-115.965	124.058	0.0	-1447.602	181.171	118.455	0.0	2158.328

We invert the matrix using the MATHCAD programme and call the inverted matrix KC^{-1}

$$KC^{-1} = \begin{bmatrix} 0.0032 & 0.0038 & 1.7449 \bullet 10^{-5} & 0.0035 & 0.0013 & 7.1397 \bullet 10^{-4} & 0.0015 & 0.0035 & -5.8252 \bullet 10^{-5} & 4.9344 \bullet 10^{-4} \\ 0.0038 & 0.0178 & -0.0035 & 0.0169 & 0.0032 & 0.0047 & -1.3862 \bullet 10^{-5} & 0.0171 & -0.0032 & 0.0048 \\ 1.7449 \bullet 10^{-5} & -0.0035 & 0.0032 & -0.0038 & -5.8252 \bullet 10^{-5} & -4.9344 \bullet 10^{-4} & 0.0015 & -0.0035 & 0.0013 & -7.1397 \bullet 10^{-4} \\ 0.0035 & 0.0169 & -0.0038 & 0.0178 & 0.0032 & 0.0048 & 1.3862 \bullet 10^{-5} & 0.0171 & -0.0032 & 0.0047 \\ 0.0013 & 0.0032 & -5.8252 \bullet 10^{-5} & 0.0032 & 0.0013 & 9.1423 \bullet 10^{-4} & 5.1552 \bullet 10^{-4} & 0.0031 & -3.5629 \bullet 10^{-4} & 8.2001 \bullet 10^{-4} \\ 7.1397 \bullet 10^{-4} & 0.0047 & -4.9344 \bullet 10^{-4} & 0.0048 & 9.1423 \bullet 10^{-4} & 0.0023 & 1.4139 \bullet 10^{-4} & 0.0047 & -8.2001 \bullet 10^{-4} & 0.0020 \\ 0.0015 & -1.3862 \bullet 10^{-5} & 0.0015 & 1.3862 \bullet 10^{-5} & 5.1552 \bullet 10^{-4} & 1.4139 \bullet 10^{-4} & 0.0018 & 0.0000 & 5.1552 \bullet 10^{-4} & -1.4139 \bullet 10^{-4} \\ 0.0035 & 0.0171 & -0.0035 & 0.0171 & 0.0031 & 0.0047 & 0.0000 & 0.0169 & -0.0031 & 0.0047 \\ -5.8252 \bullet 10^{-5} & -0.0032 & 0.0013 & -0.0032 & -3.5629 \bullet 10^{-4} & -8.2001 \bullet 10^{-4} & 5.1552 \bullet 10^{-4} & -0.0031 & 0.0013 & -9.1423 \bullet 10^{-4} \\ 4.9344 \bullet 10^{-4} & 0.0048 & -7.1397 \bullet 10^{-4} & 0.0047 & 8.2001 \bullet 10^{-4} & 0.0020 & -1.4139 \bullet 10^{-4} & 0.0047 & -9.1423 \bullet 10^{-4} & 0.0023 \end{bmatrix}$$

We extract the 10 unknown nodal displacements by multiplying this inverted matrix, which we call S, by the reduced nodal force vector which we call Z

$$Z := \begin{bmatrix} 1 \\ 0 \\ 1 \\ 0 \\ 0 \\ 0 \\ 0 \\ 0 \\ 0 \\ 0 \end{bmatrix}$$

$$S := KC^{-1}$$

So the ten unknown displacements are, in metres:

$$S \cdot Z = \begin{bmatrix} 0.003177 \\ 2.144592 \bullet 10^{-4} \\ 0.003177 \\ -2.144592 \bullet 10^{-4} \\ 0.001193 \\ 2.205301 \bullet 10^{-4} \\ 0.002974 \\ 0.000000 \\ 0.001193 \\ -2.205301 \bullet 10^{-4} \end{bmatrix} = \begin{bmatrix} u_2 \\ v_2 \\ u_3 \\ v_3 \\ u_5 \\ v_5 \\ u_6 \\ v_6 \\ u_7 \\ v_7 \end{bmatrix}$$

We add to this set of nodal displacements the known zero displacements at nodes 1, 4 and 8

169

and we call this complete nodal displacement vector Q. We place alongside it the corresponding set of nodal displacements $Q4$ calculated in the previous example using four Gauss points. There is close, but not exact, agreement between the two sets of data.

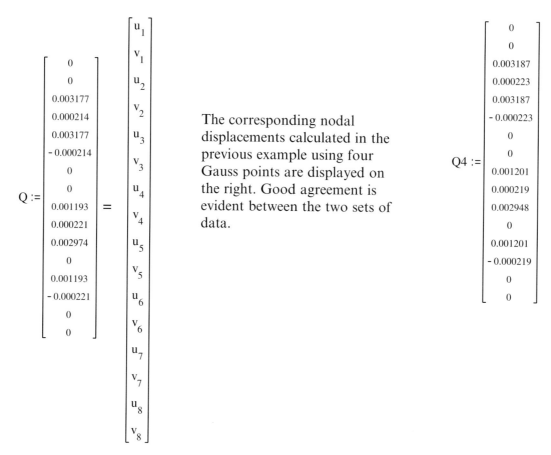

$$Q := \begin{bmatrix} 0 \\ 0 \\ 0.003177 \\ 0.000214 \\ 0.003177 \\ -0.000214 \\ 0 \\ 0 \\ 0.001193 \\ 0.000221 \\ 0.002974 \\ 0 \\ 0.001193 \\ -0.000221 \\ 0 \\ 0 \end{bmatrix} = \begin{bmatrix} u_1 \\ v_1 \\ u_2 \\ v_2 \\ u_3 \\ v_3 \\ u_4 \\ v_4 \\ u_5 \\ v_5 \\ u_6 \\ v_6 \\ u_7 \\ v_7 \\ u_8 \\ v_8 \end{bmatrix}$$

The corresponding nodal displacements calculated in the previous example using four Gauss points are displayed on the right. Good agreement is evident between the two sets of data.

$$Q4 := \begin{bmatrix} 0 \\ 0 \\ 0.003187 \\ 0.000223 \\ 0.003187 \\ -0.000223 \\ 0 \\ 0 \\ 0.001201 \\ 0.000219 \\ 0.002948 \\ 0 \\ 0.001201 \\ -0.000219 \\ 0 \\ 0 \end{bmatrix}$$

8.2.5 ARE THE NODAL FORCES IN STATIC EQUILIBRIUM?

As in the previous case we make a check on the equilibrium of the forces applied to the element by exploring if the resultant horizontal force is zero. We extract the complete set of nodal applied forces by multiplying the uncondensed stiffness matrix KR by the nodal displacement vector Q. The result is

$$KR \cdot Q = \begin{bmatrix} -0.02200 \\ -0.12442 \\ 1.00026 \\ -3.65403 \cdot 10^{-4} \\ 1.00019 \\ 3.55030 \cdot 10^{-4} \\ -0.02192 \\ 0.12441 \\ -3.08764 \cdot 10^{-6} \\ 1.49176 \cdot 10^{-3} \\ -7.83655 \cdot 10^{-4} \\ 5.44030 \cdot 10^{-6} \\ 7.06188 \cdot 10^{-5} \\ -1.41079 \cdot 10^{-3} \\ -1.95595 \\ -4.91125 \cdot 10^{-5} \end{bmatrix}$$

The horizontal force components (in the direction of x) in this force vector are in rows 1, 3, 5, 7, 9, 11, 13 and 15.

These forces should add to zero if they are in static equilibrium. The sum is

$$-0.022 + 1.00026 + 1.00019 - 0.02192 - 0.000003 - .000783$$
$$-.0000706 - 1.95595 = -0.00028 \text{ MN}$$

This is sufficiently close to zero to persuade us that the solution preserves static equilibrium and our confidence in the results is enhanced.

8.2.6 Calculation of the Gauss Point Stresses

We are now in a position to calculate the stresses at the Gauss points. We use the relation

$$\sigma = [D][B][Q] \text{ for each Gauss point}$$

We have already set up the $[B]$ matrices, for example that at Gauss point a is Ba. The results are listed below.

Recall that the three terms in each stress vector are

$$\begin{bmatrix} \sigma_{xx} \\ \sigma_{yy} \\ \sigma_{xy} \end{bmatrix}$$

STRESSES AT GAUSS POINTS

Gauss point a

$$D \cdot Ba \cdot Q = \begin{bmatrix} 69.69 \\ 13.94 \\ 5.02 \end{bmatrix} \begin{bmatrix} \sigma_{xx} \\ \sigma_{yy} \\ \sigma_{xy} \end{bmatrix}$$

point b

$$D \cdot Bb \cdot Q = \begin{bmatrix} 112.41 \\ 3.09 \\ -3.89 \end{bmatrix} \begin{bmatrix} \sigma_{xx} \\ \sigma_{yy} \\ \sigma_{xy} \end{bmatrix}$$

point c

$$D \cdot Bc \cdot Q = \begin{bmatrix} 160.36 \\ -5.16 \\ -21.72 \end{bmatrix} \begin{bmatrix} \sigma_{xx} \\ \sigma_{yy} \\ \sigma_{xy} \end{bmatrix}$$

point d

$$D \cdot Bd \cdot Q = \begin{bmatrix} 64.28 \\ 12.32 \\ 0.00 \end{bmatrix} \begin{bmatrix} \sigma_{xx} \\ \sigma_{yy} \\ \sigma_{xy} \end{bmatrix}$$

point e

$$D \cdot Be \cdot Q = \begin{bmatrix} 103.14 \\ 0.31 \\ 0.00 \end{bmatrix} \begin{bmatrix} \sigma_{xx} \\ \sigma_{yy} \\ \sigma_{xy} \end{bmatrix}$$

point f

$$D \cdot Bf \cdot Q = \begin{bmatrix} 142.33 \\ -10.57 \\ 0.00 \end{bmatrix} \begin{bmatrix} \sigma_{xx} \\ \sigma_{yy} \\ \sigma_{xy} \end{bmatrix}$$

point g

$$D \cdot Bg \cdot Q = \begin{bmatrix} 69.70 \\ 13.94 \\ -5.01 \end{bmatrix} \begin{bmatrix} \sigma_{xx} \\ \sigma_{yy} \\ \sigma_{xy} \end{bmatrix}$$

point h

$$D \cdot Bh \cdot Q = \begin{bmatrix} 112.41 \\ 3.09 \\ 3.89 \end{bmatrix} \begin{bmatrix} \sigma_{xx} \\ \sigma_{yy} \\ \sigma_{xy} \end{bmatrix}$$

point k

$$D \cdot Bk \cdot Q = \begin{bmatrix} 160.36 \\ -5.16 \\ 21.72 \end{bmatrix} \begin{bmatrix} \sigma_{xx} \\ \sigma_{yy} \\ \sigma_{xy} \end{bmatrix}$$

Note that, as we would expect from the symmetry of the element, the direct stresses σ_{xx} and σ_{yy} are equal at the following pairs of Gauss points a and g, c and k, b and h. The shear stresses σ_{xy} are equal but of opposite sign at these pairs of location.

8.2.7 Calculation of the Nodal Stresses

Finally we calculate the nodal stresses. This section of the calculation is identical to that in the four-Gauss-point example. The values of the A, G and B matrices are identical because we have used exactly the same eight nodal coordinates in the two examples. However we have reproduced the A and G matrices to emphasise that equality. The matrices are designated $A1$ and $G1$ for node 1, $A2$ and $G2$ for node 2 and so on.

The Nodal Stresses

NODE 1

$$A1 := 1.33333 \cdot \begin{bmatrix} .75 & -.25 & 0 & 0 \\ 0 & 0 & 0 & 1 \\ 0 & 1 & .75 & -.25 \end{bmatrix}$$

$$G1 := \begin{bmatrix} -1.5 & 0 & -.5 & 0 & 0 & 0 & 0 & 0 & 2 & 0 & 0 & 0 & 0 & 0 & 0 \\ -1.5 & 0 & 0 & 0 & 0 & 0 & -.5 & 0 & 0 & 0 & 0 & 0 & 0 & 2 & 0 \\ 0 & -1.5 & 0 & -.5 & 0 & 0 & 0 & 0 & 0 & 2 & 0 & 0 & 0 & 0 & 0 \\ 0 & -1.5 & 0 & 0 & 0 & 0 & 0 & -.5 & 0 & 0 & 0 & 0 & 0 & 0 & 2 \end{bmatrix}$$

$$B1 := A1 \cdot G1$$

$$D \cdot B1 \cdot Q = \begin{bmatrix} 60.73 \\ 18.22 \\ 8.93 \end{bmatrix}$$

NODE 2

$$A2 := 4 \cdot \begin{bmatrix} .25 & -.25 & 0 & 0 \\ 0 & 0 & 0 & 1 \\ 0 & 1 & .25 & -.25 \end{bmatrix}$$

$$G2 := \begin{bmatrix} .5 & 0 & 1.5 & 0 & 0 & 0 & 0 & 0 & -2 & 0 & 0 & 0 & 0 & 0 & 0 \\ 0 & 0 & -1.5 & 0 & -.5 & 0 & 0 & 0 & 0 & 0 & 2 & 0 & 0 & 0 & 0 \\ 0 & .5 & 0 & 1.5 & 0 & 0 & 0 & 0 & 0 & -2 & 0 & 0 & 0 & 0 & 0 \\ 0 & 0 & 0 & -1.5 & 0 & -.5 & 0 & 0 & 0 & 0 & 0 & 2 & 0 & 0 & 0 \end{bmatrix}$$

$$B2 := A2 \cdot G2$$

$$D \cdot B2 \cdot Q = \begin{bmatrix} 192.56 \\ -1.55 \\ -40.80 \end{bmatrix}$$

NODE 3

$$A3 := 4 \cdot \begin{bmatrix} .25 & .25 & 0 & 0 \\ 0 & 0 & 0 & 1 \\ 0 & 1 & .25 & .25 \end{bmatrix}$$

$$G3 := \begin{bmatrix} 0 & 0 & 0 & 0 & 1.5 & 0 & .5 & 0 & 0 & 0 & 0 & 0 & -2 & 0 & 0 & 0 \\ 0 & 0 & .5 & 0 & 1.5 & 0 & 0 & 0 & 0 & 0 & -2 & 0 & 0 & 0 & 0 & 0 \\ 0 & 0 & 0 & 0 & 0 & 1.5 & 0 & .5 & 0 & 0 & 0 & 0 & 0 & -2 & 0 & 0 \\ 0 & 0 & 0 & .5 & 0 & 1.5 & 0 & 0 & 0 & 0 & 0 & -2 & 0 & 0 & 0 & 0 \end{bmatrix}$$

$$B3 := A3 \cdot G3$$

$$D \cdot B3 \cdot Q = \begin{bmatrix} 192.56 \\ -1.55 \\ 40.80 \end{bmatrix}$$

NODE4

$$A4 := 1.33333 \cdot \begin{bmatrix} .75 & .25 & 0 & 0 \\ 0 & 0 & 0 & 1 \\ 0 & 1 & .75 & .25 \end{bmatrix}$$

$$G4 := \begin{bmatrix} 0 & 0 & 0 & 0 & -.5 & 0 & -1.5 & 0 & 0 & 0 & 0 & 0 & 2 & 0 & 0 & 0 \\ .5 & 0 & 0 & 0 & 0 & 0 & 1.5 & 0 & 0 & 0 & 0 & 0 & 0 & 0 & -2 & 0 \\ 0 & 0 & 0 & 0 & 0 & -.5 & 0 & -1.5 & 0 & 0 & 0 & 0 & 0 & 2 & 0 & 0 \\ 0 & .5 & 0 & 0 & 0 & 0 & 0 & 1.5 & 0 & 0 & 0 & 0 & 0 & 0 & 0 & -2 \end{bmatrix}$$

$$B4 := A4 \cdot G4$$

$$D \cdot B4 \cdot Q = \begin{bmatrix} 60.73 \\ 18.22 \\ -8.93 \end{bmatrix}$$

173

NODE 5

$$A5 := 2 \cdot \begin{bmatrix} .5 & -.25 & 0 & 0 \\ 0 & 0 & 0 & 1 \\ 0 & 1 & .5 & -.25 \end{bmatrix}$$

$$G5 := \begin{bmatrix} -.5 & 0 & .5 & 0 & 0 & 0 & 0 & 0 & 0 & 0 & 0 & 0 & 0 & 0 & 0 & 0 \\ -.5 & 0 & -.5 & 0 & -.5 & 0 & -.5 & 0 & -.5 & 0 & 1 & 0 & .5 & 0 & 1 & 0 \\ 0 & -.5 & 0 & .5 & 0 & 0 & 0 & 0 & 0 & 0 & 0 & 0 & 0 & 0 & 0 & 0 \\ 0 & -.5 & 0 & -.5 & 0 & -.5 & 0 & -.5 & 0 & -.5 & 0 & 1 & 0 & .5 & 0 & 1 \end{bmatrix}$$

$$B5 := A5 \cdot G5$$

$$D \cdot B5 \cdot Q = \begin{bmatrix} 118.60 \\ 4.95 \\ -5.02 \end{bmatrix}$$

NODE 6

$$A6 := 4 \cdot \begin{bmatrix} .25 & 0 & 0 & 0 \\ 0 & 0 & 0 & 1 \\ 0 & 1 & .25 & 0 \end{bmatrix}$$

$$G6 := \begin{bmatrix} .5 & 0 & .5 & 0 & .5 & 0 & .5 & 0 & -1 & 0 & .5 & 0 & -1 & 0 & -.5 & 0 \\ 0 & 0 & -.5 & 0 & .5 & 0 & 0 & 0 & 0 & 0 & 0 & 0 & 0 & 0 & 0 & 0 \\ 0 & .5 & 0 & .5 & 0 & .5 & 0 & .5 & 0 & -1 & 0 & .5 & 0 & -1 & 0 & -.5 \\ 0 & 0 & 0 & -.5 & 0 & .5 & 0 & 0 & 0 & 0 & 0 & 0 & 0 & 0 & 0 & 0 \end{bmatrix}$$

$$B6 := A6 \cdot G6$$

$$D \cdot B6 \cdot Q = \begin{bmatrix} 153.91 \\ -13.14 \\ 0.00 \end{bmatrix}$$

174

NODE 7

$$A7 := 2 \cdot \begin{bmatrix} .5 & .25 & 0 & 0 \\ 0 & 0 & 0 & 1 \\ 0 & 1 & .5 & .25 \end{bmatrix}$$

$$G7 := \begin{bmatrix} 0 & 0 & 0 & 0 & .5 & 0 & -.5 & 0 & 0 & 0 & 0 & 0 & 0 & 0 & 0 & 0 \\ .5 & 0 & .5 & 0 & .5 & 0 & .5 & 0 & -.5 & 0 & -1 & 0 & .5 & 0 & -1 & 0 \\ 0 & 0 & 0 & 0 & 0 & .5 & 0 & -.5 & 0 & 0 & 0 & 0 & 0 & 0 & 0 & 0 \\ 0 & .5 & 0 & .5 & 0 & .5 & 0 & .5 & 0 & -.5 & 0 & -1 & 0 & .5 & 0 & -1 \end{bmatrix}$$

$$B7 := A7 \cdot G7$$

$$D \cdot B7 \cdot Q = \begin{bmatrix} 118.60 \\ 4.95 \\ 5.02 \end{bmatrix}$$

NODE 8

$$A8 := 1.33333 \cdot \begin{bmatrix} .75 & 0 & 0 & 0 \\ 0 & 0 & 0 & 1 \\ 0 & 1 & .75 & 0 \end{bmatrix}$$

$$G8 := \begin{bmatrix} -.5 & 0 & -.5 & 0 & -.5 & 0 & -.5 & 0 & 1 & 0 & .5 & 0 & 1 & 0 & -.5 & 0 \\ -.5 & 0 & 0 & 0 & 0 & 0 & .5 & 0 & 0 & 0 & 0 & 0 & 0 & 0 & 0 & 0 \\ 0 & -.5 & 0 & -.5 & 0 & -.5 & 0 & -.5 & 0 & 1 & 0 & .5 & 0 & 1 & 0 & -.5 \\ 0 & -.5 & 0 & 0 & 0 & 0 & 0 & .5 & 0 & 0 & 0 & 0 & 0 & 0 & 0 & 0 \end{bmatrix}$$

$$B8 := A8 \cdot G8$$

$$D \cdot B8 \cdot Q = \begin{bmatrix} 53.00 \\ 15.90 \\ 0.00 \end{bmatrix}$$

CENTROID

$$a9 := 2.0 \cdot \begin{bmatrix} .5 & 0 & 0 & 0 \\ 0 & 0 & 0 & 1 \\ 0 & 1 & .5 & 0 \end{bmatrix}$$

$$g9 := \begin{bmatrix} 0 & 0 & 0 & 0 & 0 & 0 & 0 & 0 & 0 & 0 & .5 & 0 & 0 & 0 & -.5 & 0 \\ 0 & 0 & 0 & 0 & 0 & 0 & 0 & 0 & -.5 & 0 & 0 & 0 & .5 & 0 & 0 & 0 \\ 0 & 0 & 0 & 0 & 0 & 0 & 0 & 0 & 0 & 0 & 0 & .5 & 0 & 0 & 0 & -.5 \\ 0 & 0 & 0 & 0 & 0 & 0 & 0 & 0 & -.5 & 0 & 0 & 0 & .5 & 0 & 0 & 0 \end{bmatrix}$$

$$a9 \cdot g9 = \begin{bmatrix} 0.000 & 0.000 & 0.000 & 0.000 & 0.000 & 0.000 & 0.000 & 0.000 & 0.000 & 0.000 & 0.500 & 0.000 & 0.000 & 0.000 & -0.500 & 0.000 \\ 0.000 & 0.000 & 0.000 & 0.000 & 0.000 & 0.000 & 0.000 & 0.000 & 0.000 & -1.000 & 0.000 & 0.000 & 0.000 & 1.000 & 0.000 & 0.000 \\ 0.000 & 0.000 & 0.000 & 0.000 & 0.000 & 0.000 & 0.000 & 0.000 & -1.000 & 0.000 & 0.000 & 0.500 & 1.000 & 0.000 & 0.000 & -0.500 \end{bmatrix}$$

$$B9 := a9 \cdot g9$$

$$D \cdot B9 \cdot Q = \begin{bmatrix} 103.14 \\ 0.31 \\ 0.00 \end{bmatrix}$$

8.3 COMPARISON OF THE NODAL STRESSES FOR 4-POINT AND 9-POINT INTEGRATION

In the table below, we display the values of σ_{xx} and σ_{yy} at selected nodes for the two calculations – with 4 and with 9 Gauss points.

It is evident that the two sets of data agree quite well and that the considerable additional computation involved in the 9-point case is hardly justified by the modest difference in numerical output. This will be an even more powerful conclusion with a meaningful engineering model that may have hundreds of elements.

The combination of the number of nodes and the number of integration points that give the optimum solution is a complex issue. However the experts appear to agree that, for the 8-node element that we have chosen, four integration points are recommended. Our calculations are consistent with that advice.

Node	Stress	Stress Values Calculated using	
		4 Gauss points	9 Gauss points
1	σ_{xx}	61.57	60.73
	σ_{yy}	18.47	18.21
2	σ_{xx}	197.14	192.56
	σ_{yy}	−2.67	−1.55
5	σ_{xx}	120.44	118.60
	σ_{yy}	5.78	4.95
6	σ_{xx}	151.65	153.91
	σ_{yy}	−16.32	−13.14
8	σ_{xx}	52.47	53.00
	σ_{yy}	15.74	15.90
Centroid	σ_{xx}	102.24	103.14
	σ_{yy}	0.32	0.31

8.4 COMPARISON WITH THE RESULTS OF RUNNING THE PROBLEM ON COMMERCIAL FINITE ELEMENT PACKAGES

It is instructive to run this loaded tapered sheet problem on commercially available finite element packages. We quote the results from three such – LUSAS, ABAQUS and IDEAS.

8.4.1 EIGHT NODES AND FOUR INTEGRATION POINTS

We first compare results from analyses that use an eight-node element and four integration points.

All the packages give values close to those we have calculated for the nodal displacements and for the stresses at the Gauss points. A comparison of the Gauss point stresses is collected in Table 8.4.1. The stress unit is MPa.

Table 8.4.1

Gauss Point	Stress	Our Value	LUSAS	ABAQUS	IDEAS
a	σ_{xx}	77.55	77.60	77.68	77.60
a	σ_{yy}	10.49	10.52	10.50	10.01
b	σ_{xx}	140.61	140.58	140.90	140.58
b	σ_{yy}	−6.05	−6.09	−6.06	−5.42
c	σ_{xx}	140.61	140.58	140.90	140.58
c	σ_{yy}	−6.05	−6.09	−6.06	−5.42
d	σ_{xx}	77.55	77.60	77.68	77.60
d	σ_{yy}	10.49	10.52	10.50	10.01

All the packages give values that agree among themselves and are very close to our calculated stresses.

In contrast, the values that the packages return for the nodal stresses differ significantly from our calculated values. Table 8.4.2 presents the data

Table 8.4.2

Node	Stress	Our value	LUSAS	ABAQUS	IDEAS	Our values with 4-node extrapolation
1	σ_{xx}	61.57	54.54	54.55	54.55	54.47
1	σ_{yy}	18.47	16.60	16.56	15.66	16.61
2	σ_{xx}	197.14	163.6	164.0	163.6	163.6
2	σ_{yy}	−2.67	−12.16	−12.12	−11.07	−12.25
5	σ_{xx}	120.44	109.1	109.3	109.1	109.1
5	σ_{yy}	5.78	2.22	2.22	2.30	2.22
6	σ_{xx}	151.65	163.6	164.0	163.6	163.7
6	σ_{yy}	−16.32	−12.16	−12.12	−11.07	−12.19
8	σ_{xx}	52.47	54.54	54.55	54.55	54.47
8	σ_{yy}	15.74	16.60	16.56	15.66	16.61

Evidently the results from all three packages agree among themselves but differ significantly from our stress values.

After a lengthy discussion with LUSAS it emerged that the reason for the difference between our nodal stresses and theirs is that, although they use quadratic shape functions to calculate the

177

displacements at the eight nodes and the stresses at the four Gauss points, they use linear shape functions to extrapolate from the Gauss point stresses to the nodal stresses.

They do this in the following way. They set up a fictitious 4-node square serendipity element which has the four Gauss points at the corners. These points are given the co-ordinates $(\pm 1, \pm 1)$ – see Fig. 8.4.1. The scale of the element has been increased by the factor $\sqrt{3}$ because in the original element the co-ordinates of the Gauss points were $\left(\pm\frac{1}{\sqrt{3}}, \pm\frac{1}{\sqrt{3}}\right)$ not $(\pm 1, \pm 1)$ as they now are.

The element is now extended to form a larger square on which the 8 nodes are located and the relative positions of the Gauss points and the nodes are preserved. This is achieved if the co-ordinates of the nodes are

Node	ξ	η
1	$-\sqrt{3}$	$-\sqrt{3}$
2	$+\sqrt{3}$	$-\sqrt{3}$
3	$+\sqrt{3}$	$+\sqrt{3}$
4	$-\sqrt{3}$	$+\sqrt{3}$
5	0	$-\sqrt{3}$
6	$+\sqrt{3}$	0
7	0	$+\sqrt{3}$
8	$-\sqrt{3}$	0

The nodal stresses are now derived from the Gauss point stresses using an iso-parametric relation. We introduced this relation in Section 7.2.1 (eqn 7.2.2). The essence of this relation is that if we know the values of any quantity ϕ at the nodes, for example the nodal co-ordinate x or the nodal displacement u, we can get the value of that quantity at any point in the element prescribed by its master element co-ordinates (ξ, η). The relation that achieves this end is

$$\phi = N_1 \cdot \phi_1 + N_2 \cdot \phi_2 + N_3 \cdot \phi_3 + N_4 \cdot \phi_4$$

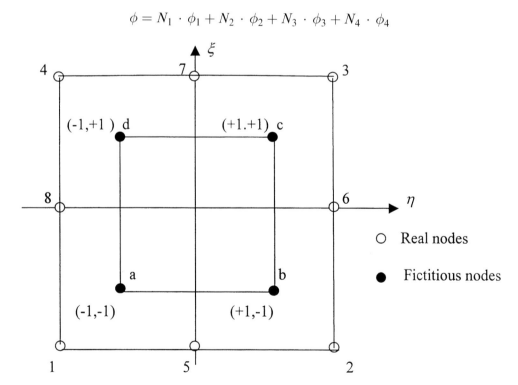

Fig. 8.4.1   The four-node serendipity fictitious element.

Where ϕ_1 to ϕ_4 are the nodal values and N_1 to N_4 are the four-node linear shape functions. We have seen that these latter are functions of the master element co-ordinates (ξ, η). The relation for ϕ is called an iso-parametric relation because the same type of equation with the same shape functions gives both the real co-ordinate and the elastic displacement at an internal point in the element.

We apply this iso-parametric relation to the present problem as follows. In the fictitious element we regard the re-positioned Gauss points a, b, c and d as the nodes at which a particular stress component has values σ_a, σ_b, σ_c and σ_d respectively. The stress σ at any point in the extended element at the point (ξ, η) is then given by

$$\sigma = N_a\sigma_a + N_b\sigma_b + N_c\sigma_c + N_d\sigma_d$$

where the linear shape functions are

$$N_a = \tfrac{1}{4}(1 - \xi)(1 - \eta) \qquad N_b = \tfrac{1}{4}(1 + \xi)(1 - \eta)$$
$$N_c = \tfrac{1}{4}(1 + \xi)(1 + \eta) \qquad N_d = \tfrac{1}{4}(1 - \xi)(1 + \eta)$$

Note that each shape function should be unity at its node so, for example, node b is at $(1. - 1)$ and

$$N_b = \tfrac{1}{4}(1 + 1)(1 - \{-1\}) = 1 \quad \text{which checks.}$$

Let us see what nodal stresses this extrapolation gives from our Gauss point stresses. We look first at σ_{xx}, the direct stress in the x direction. Our Gauss point values for this stress component are

Gauss point	σ_{xx} (Mpa)
a	77.55
b	140.61
c	140.61
d	77.55

We determine now the value of σ_{xx}^1 at node 1 on the periphery of the extended element the co-ordinates of which are $\xi = -\sqrt{3}$, $\eta = -\sqrt{3}$ so

$$\sigma_{xx}^1 = N_a\sigma_{xx}^a + N_b\sigma_{xx}^b + N_c\sigma_{xx}^c + N_d\sigma_{xx}^d$$

substituting the nodal co-ordinates for node 1 in the shape function expressions yields

$$\sigma_{xx}^1 = [1/4(1 + \sqrt{3})(1 + \sqrt{3})]77.55 + [1/4(1 - \sqrt{3})(1 + \sqrt{3})]140.61$$
$$+ [1/4(1 - \sqrt{3})(1 - \sqrt{3})]140.61 + [1/4(1 + \sqrt{3})(1 - \sqrt{3})]77.55$$
$$= 54.47 \, \text{Mpa}$$

The LUSAS value is 54.54 Mpa so agreement is now good. The other seven values of σ_{xx} can be calculated in a similar way.

We can also calculate the values of σ_{yy} by a similar procedure. The corresponding extrapolation equation is

$$\sigma_{yy} = \sigma_{yy}^a N_a + \sigma_{yy}^b N_b + \sigma_{yy}^c N_c + \sigma_{yy}^d N_d$$

The Gauss point values of σ_{yy} are

Gauss point	σ_{yy}
a	10.49
b	−6.05
c	−6.05
d	10.49

As an example we calculate the value of σ_{yy}^5 at node 5 at which the coordinates are $\xi = 0$, $\eta = -\sqrt{3}$

$$\sigma_{yy}^5 = [1/4(1 + \sqrt{3})]10.49 - [1/4](1 + \sqrt{3})]6.05 - [1/4(1 - 3)]6.05 + [1/4(1 - \sqrt{3})]10.49$$
$$= 2.22 \, \text{Mpa}$$

The LUSAS value is 2.22 Mpa.

The stress values produced by using the linear shape function extrapolation of our Gauss point stresses are listed in the last column of Table 8.4.2. It is evident that by using this extrapolation our nodal stress values are reconciled with those delivered by the finite element packages. ABAQUS performs the extrapolation in the same way as does LUSAS and it must be presumed that IDEAS does also.

8.4.2 EIGHT NODES AND NINE INTEGRATION POINTS

The ABAQUS package, using the element type designated CPS8, gives values for the nodal displacements and for the Gauss point stresses that are in good agreement with the results of our calculations. The following Table 8.4.3 compares a selection of Gauss point stress values to indicate the level of agreement.

Table 8.4.3

Gauss Point	Stress	Stresses at the Gauss points (Mpa) Our Value	ABAQUS Value
a	σ_{xx}	69.69	69.73
	σ_{yy}	13.94	13.96
	σ_{xy}	5.02	5.01
b	σ_{xx}	112.41	112.6
	σ_{yy}	3.09	3.16
	σ_{xy}	−3.89	−3.87
c	σ_{xx}	160.36	160.7
	σ_{yy}	−5.16	−5.16
	σ_{xy}	−21.72	−21.66
d	σ_{xx}	64.28	64.35
	σ_{yy}	12.32	12.34
	σ_{xy}	0	0
e	σ_{xx}	103.14	103.4
	σ_{yy}	0.31	−0.38
	σ_{xy}	0	0
f	σ_{xx}	142.33	142.7
	σ_{yy}	−10.57	−10.56
	σ_{xy}	0	0

The nodal stresses do not show quite such good agreement, but the level of agreement is better than that for the calculations with four integration points. The Table 8.4.4 compares selected nodal values for the direct stresses σ_{xx} and σ_{yy}.

The reason for the minor differences between our nodal stress values and the ABAQUS values is similar to that for the larger discrepancy between the nodal values in the example using four integration points.

For the CPS8 element that was used to secure the quoted data, ABAQUS use an extrapolation to get the nodal stresses from the Gauss point stresses that is similar to that which is used for the four-Gauss-point element that we explained in the previous section. The Gauss points are made the nodes of a square 'fictitious element' and the element is extended beyond these nodes in the same fashion as was done for the four-Gauss-point element. In the case of the 9-Gauss point element, this extension is by a factor of $1/0.7746 = 1.291$ and the real nodes lie on the edge of this extended square. The 'fictitious element' is drawn in Fig. 8.4.2 and compares with the four-Gauss-point equivalent of Fig. 8.4.1.

Since there are nine nodes it is necessary to use shape functions appropriate to a nine-node element so there must be nine shape functions. The appropriate shape functions in this case are a second order set called Lagrangian, which we reproduce here for reference.

The Nine-node Lagrangian Shape Functions

$$N_a = \frac{\xi\eta}{4}(1-\xi)(1-\eta) \qquad\qquad N_f = \frac{\xi}{2}(1+\xi)(1-\eta^2)$$

$$N_b = -\frac{\eta}{2}(1-\eta)(1-\xi^2) \qquad\qquad N_g = -\frac{\xi\eta}{4}(1-\xi)(1+\eta)$$

$$N_c = -\frac{\xi\eta}{4}(1+\xi)(1-\eta) \qquad\qquad N_h = \frac{\eta}{2}(1+\eta)(1-\xi^2)$$

$$N_d = -\frac{\xi}{2}(1-\xi)(1-\eta^2) \qquad\qquad N_k = \frac{\xi\eta}{4}(1+\xi)(1+\eta)$$

$$N_e = (1-\xi^2)(1-\eta^2)$$

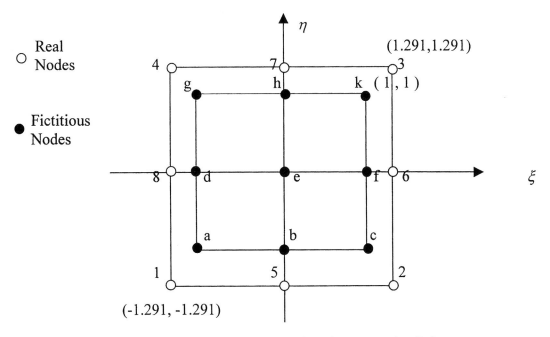

Fig. 8.4.2 The 'Fictitious Element' for Nine Integration Points.

These shape functions are used in an iso-parametric form of equation similar to that for the four-point extrapolation. In this case there are nine terms so, if the values of a particular stress component at the Gauss points are σ_a to σ_k and the nine shape functions are N_a to N_k then the extrapolated value of the stress component σ at any point in the "fictitious element" is given by

$$\sigma = \sigma_a N_a + \sigma_b N_b + \sigma_c N_c + \sigma_d N_d + \sigma_e N_e + \sigma_f N_f + \sigma_g N_g + \sigma_h N_h + \sigma_k N_k$$

As an example of the application of this expression we outline the procedure for calculating the value of σ_{xx} at node 1. The co-ordinates of this node in the fictitious element are

$$\xi = -1.291 \quad \text{and} \quad \eta = -1.291.$$

We substitute these values in the expressions for the nine shape functions, which are, of course, functions of ξ and η. Then we put these values together with the Gauss point values of σ_{xx} in the following iso-parametric relation

$$\sigma_{xx}^1 = \sigma_{xx}^a N_a + \sigma_{xx}^b N_b + \sigma_{xx}^c N_c + \sigma_{xx}^d N_d + \sigma_{xx}^e N_e + \sigma_{xx}^f N_f + \sigma_{xx}^g N_g + \sigma_{xx}^h N_h + \sigma_{xx}^k N_k$$

where σ_{xx}^a is the value of σ_{xx} at Gauss point a and so on.

If we put in the appropriate stress values we find that

$$\sigma_{xx}^1 = 61.67 \, \text{Mpa}$$

which is exactly the same as the ABAQUS value of 61.67 Mpa.

A number of nodal values of σ_{xx} and σ_{yy} have been calculated from our Gauss point stress values in this way. The results are listed in the last column of Table 8.4.4. The values all agree excellently with the ABAQUS values.

Table 8.4.4

Node	Stress	A Comparison of Nodal Stress Values		
		Our value	ABAQUS value	Our value using ABAQUS extrapolation
1	σ_{xx}	60.73	61.67	61.67
	σ_{yy}	18.22	18.61	18.62
2	σ_{xx}	192.56	190.0	189.6
	σ_{yy}	−1.55	−2.82	−2.77
5	σ_{xx}	118.60	118.8	118.6
	σ_{yy}	4.95	5.01	4.94
6	σ_{xx}	153.91	154.1	153.8
	σ_{yy}	−13.14	−13.55	−13.52
8	σ_{xx}	53.00	53.05	53.03
	σ_{yy}	15.90	16.02	16.03

8.4.3 A Final Comment on the Comparison between Our Data and the Output from the Commercial Packages

The Gauss point stresses are the most accurate calculated values of the stresses in the element. The good agreement between these stresses which we have calculated and those derived from the commercial packages emphasises that the same method of calculation has been used in both cases.

The differences between our values of the nodal stresses and those calculated by the commercial packages reveal that different methods have been used to derive the nodal stresses.

What we have done is to use the basic finite element relation between the nodal displacements $\{u\}$ and the elastic strain vector $\{\varepsilon\}$

$$\{\varepsilon\} = [B]\{u\}.$$

If the co-ordinates of a particular node are put into the matrix $[B]$ and the nodal displacements $\{u\}$ are known, then the strain vector at the node can be derived. The stresses at the node then follow from the basic elastic stress–strain relation

$$\{\sigma\} = [D]\{\varepsilon\}$$
$$= [D][B]\{u\}$$

where $[D]$ is the matrix of elastic constants.

On the other hand, the commercial packages use an iso-parametric form of relation to extrapolate the Gauss point stresses to the nodes. We have explained that, to this end, the Gauss points are made 'fictitious nodes' in a master element which is extended beyond the fictitious nodes by a factor that depends on the position of the Gauss points in the original element. The value of a stress component at a node is then obtained by the relation

$$\sigma = \sigma_a N_a + \sigma_b N_b + \sigma_c N_c + \sigma_d N_d + \sigma_e N_e + \cdots$$

where σ_a etc. are the original Gauss point stresses and the Ns are the shape functions. The number of shape functions must be the same as the number of the original integration points. The value of the stress σ at a node is obtained by substituting the nodal (ξ, η) co-ordinates into the shape functions and substituting the values in the above equation. The nodal co-ordinates can exceed unity because the real nodes lie outside the fictitious nodes which themselves lie on a 2×2 square.

The moral to be drawn from these rather laborious numerical comparisons is that, in using a particular finite element package, it is important to be informed about the basis of important steps in the calculations. It should be possible to secure this information from the package handbooks, but unfortunately these are often over-terse and provide inadequately transparent accounts of their computational philosophy.

CHAPTER 9   A CAUTIONARY EPILOGUE

The commercial finite element packages present an irresistible computational fascination to those who revel in such matters. However, it must be appreciated that the solutions generated by these packages are approximate. The most secure way of assessing the degree of error in finite element stress analysis is to make some measurements of a small number of critical deflections and stresses on the real loaded body. To this end, extensometers, strain gauges, laser strain measurement techniques, photo-elasticity and brittle lacquer techniques all have their place. If the finite element values are in good accord with the measured quantities, then the acceptability of the solution is enhanced. Advantage can then be taken with confidence of the rich wealth of detail that the finite element package is able to provide about the stress distribution and the way it is influenced by variation of the magnitude of significant parameters. A prudent way of securing reliable finite element solution would be to declare a computer-free day once a week and use the time to make direct measurements on the system that is the object of the finite element calculations.

Metallurgical Modelling of Welding

SECOND EDITION

Øystein Grong

This book gives graduate students, engineers and researchers an in-depth insight into the field of welding metallurgy, providing a broad overview of its fundamental principles. Welding metallurgy is a fascinating interdisciplinary activity and the author takes full advantage of this to incorporate a synthesis of knowledge from diverse disciplines into the text.

In recent years, significant progress has been made in the understanding of the chemical and physical processes which take place during welding. This book brings together all the basic components necessary to reach the goal of faster process development, optimisation of process and properties, and the possibility of developing new and more weldable alloys.

Readers will learn more about physical modelling and will see how even complex materials can be treated by means of relatively simple analytical models. The text incorporates numerical examples, exercises and case studies enabling the reader to adopt a more direct theoretical approach to welding metallurgy than has been available from existing books.

The book describes a novel approach to the modelling of chemical, structural and mechanical changes in weldments, in both ferrous and non-ferrous alloys. The models treated are based on sound physical principles, but are simplified to an extent which makes them quick and easy to implement, and accessible to potential industrial users. Due to their generic nature, many of the models can be generalised to a wide range of materials and thermal processing, e.g. welding, laser surfacing, extrusion, hot forming and heat treatment.

MSc and PhD students, research scientists and engineers involved with welding in academic institutions and industry will, by reading this book, gain a thorough understanding of how the theory and models from related disciplines can be used for the best possible control of weld properties. The book has also a sufficient breadth and depth to be useful to readers outside the welding community who have a general interest in physical metallurgy and materials science.

As well as incorporating a number of corrections, the second edition includes a new chapter of exercises with solutions.

Book 677   ISBN 1 86125 036 3   Hardback
Special Student Price European Union £25 (Full Price £85)
Special Student Price Non-European Union $50 (Full Price $170)
p&p European Union £5.00/Non-EU $10.00 per order

Orders to: IOM Communications Ltd, Shelton House, Stoke Road, Shelton, Stoke-on-Trent ST4 2DR
Tel: +44 (0) 1782 202 116   Fax: +44 (0) 1782 202 421   Email: Marketing@materials.org.uk

N. American orders to: Ashgate Publishing Co., Old Post Road, Brookfield, VT 05036, USA
Tel: 802 276 3162   Fax: 802 276 3837   Email: info@ashgate.com

Credit cards accepted:

IOM Communications

Reg. Charity No. 1059475   VAT Registration No. GB 649 1646 11
IOM Communications Ltd is a wholly-owned subsidiary of The Institute of Materials

Materials for Engineering

JOHN MARTIN

This textbook presents a relatively brief overview of Materials Science, its anticipated readership being students of Structural Engineering. It is in two sections – the first characterising materials, the second considering structure/property relationships. Tabulated data in the body of the text, and the appendices, have been selected to increase the value of the book as a permanent source of reference to readers throughout their professional lives.

Contents include
The Structure of Engineering Materials
The Determination of Mechanical Properties
Metals and Alloys
Glasses and Ceramics
Organic Polymeric Materials
Composite Materials
Appendices
Index

Book 644   ISBN 1 86125 012 6   247mm x 174mm
European Union £8.50/Members £6.80
Non-European Union $17/Members $13.60
p&p European Union £5.00/Non-EU $10.00

Orders to: IOM Communications Ltd, Shelton House, Stoke Road, Shelton, Stoke-on-Trent ST4 2DR
Tel: +44 (0) 1782 202 116   Fax: +44 (0) 1782 202 421   Email: Marketing@materials.org.uk

N. American orders to: Ashgate Publishing Co., Old Post Road, Brookfield, VT 05036, USA
Tel: 802 276 3162   Fax: 802 276 3837   Email: info@ashgate.com

IOM Communications

Reg. Charity No. 1059475   VAT Registration No. GB 649 1646 11
IOM Communications Ltd is a wholly-owned subsidiary of The Institute of Materials

Concepts in the Electron Theory of Alloys

Alan Cottrell

The purpose of this book is to introduce the reader to the electron theory of metals and alloys. It explains, without the use of complex mathematics, the underlying physical ideas and how these have led to the main conclusions about the structures, compositions and properties of alloy phases. It is aimed at undergraduates and assumes that the reader has only that knowledge of the simplest aspects of the electron theory of metals which is to be found in standard undergraduate textbooks in materials science and metallurgy.

Sir Alan Cottrell FRS is a former Goldsmiths' Professor of Metallurgy at the University of Cambridge. Amongst his many publications are: *Introduction to the Modern Theory of Metals*, *Introduction to Metallurgy*, *Chemical Bonding in Transition Metal Carbides* and (as co-editor with Professor D.G. Pettifor) *Electron Theory in Alloy Design*, all of which are published by IOM Communications.

B705   ISBN 1 86125 075 4   Hbk   136pp
European Union £20/Members £16
Non-European Union $40/Members $32
p&p European Union £5.00/Non-EU $10.00 per order

Orders to: IOM Communications Ltd, Shelton House, Stoke Road, Shelton, Stoke-on-Trent ST4 2DR   Tel: +44 (0) 82 202 116   Fax: +44 (0) 1782 202 421
Email: Marketing@materials.org.uk   Internet: www.materials.org.uk

N. American orders to: Ashgate Publishing Co., Old Post Road, Brookfield, 05036 VT, USA
Tel: (802) 276 3162   Fax: (802) 276 3837   Email: info@ashgate.com

Credit cards accepted:

I O M Communications

Reg. Charity No. 1059475   VAT Registration No. GB 649 1646 11
IOM Communications Ltd is a wholly owned subsidiary of The Institute of Materials

FRACTURE MECHANICS – WORKED EXAMPLES
J. Knott and P. Withey

Fracture mechanics has become an important tool in the safe design and use of engineering components and structures. This book is aimed at those in both industry and academic institutions who require a grounding not only in the basic principles of this important field but also in the practical aspects of evaluating fracture mechanics parameters.

The reader is taken through the basics of linear elastic fracture mechanics from the Griffith Relationship, through the development of G and K, to practical engineering situations. An in-depth account is given of the procedure for making a fracture toughness measurement in accordance with BS 7448.1991. There is a major section on elastic/plastic fracture mechanics and the calculation of CTOD and J for ductile materials. The final chapter treats fatigue-growth, and its application to the calculation of lifetimes for structures containing cracks.

Book 550   ISBN 0 901462 26 6   Paperback
European Union £15/Members £12
Non-European Union $30/Members $24
p&p European Union £5.00/Non-EU $10.00

Orders to: IOM Communications Ltd, Shelton House, Stoke Road, Shelton, Stoke-on-Trent ST4 2DR
Tel: +44 (0) 1782 202 116   Fax: +44 (0) 1782 202 421   Email: Marketing@materials.org.uk

N. American orders to: Ashgate Publishing Co., Old Post Road, Brookfield, VT 05036, USA
Tel: 802 276 3162   Fax: 802 276 3837   Email: info@ashgate.com

Credit cards accepted:

IOM Communications

Reg. Charity No. 1059475   VAT Registration No. GB 649 1646 11
IOM Communications Ltd is a wholly-owned subsidiary of The Institute of Materials

Basic Corrosion Technology
for Scientists and Engineers
SECOND EDITION

Einar Mattsson

This new user-friendly survey of the corrosion of metals does not require a deep knowledge of chemistry but nevertheless acts as an invaluable compendium on the subject. The following chapter headings serve to indicate the broad scope of the author's approach:

Corrosion and its importance to the community; Basic electrochemical concepts; Basic corrosion concepts; Types of corrosion; Corrosion environments; Corrosion protection; Corrosion prevention by design; Corrosion characteristics of the most common metals in use; The methodology of corrosion investigations; Corrosion information.

Also included are useful appendices on: risk of bimetallic corrosion in different types of atmosphere; Hoover's alignment chart for determination of saturation pH (pH_s) according to the Langelier equation; some physical methods used in corrosion investigation; a selection of handbooks on corrosion; a selection of journals and abstracts periodicals for corrosion information; some regular international corrosion events.

Amply illustrated with 150 illustrations (many in colour) and with useful appendices providing further information, this book will be of value to all those encountering corrosion in their various professional activities.

Book 636   ISBN 1 86125 011 8   247mm x 174mm
European Union £20 / Members £16
Non-European Union $40 / Members $32
p&p European Union £5.00 / Non-EU $10.00

Orders to: IOM Communications Ltd, Shelton House, Stoke Road, Shelton, Stoke-on-Trent ST4 2DR
Tel: +44 (0) 1782 202 116   Fax: +44 (0) 1782 202 421   Email: Marketing@materials.org.uk

N. American orders to: Ashgate Publishing Co., Old Post Road, Brookfield, VT 05036, USA
Tel: 802 276 3162   Fax: 802 276 3837   Email: info@ashgate.com

IOM Communications

Reg. Charity No. 1059475   VAT Registration No. GB 649 1646 11
IOM Communications Ltd is a wholly-owned subsidiary of The Institute of Materials

THE SGTE CASEBOOK
Thermodynamics
at Work

Edited by
K. Hack

Contents

Book 621 ISBN 0 901716 74 X 247mm x 174mm
European Union £35/Members £28
Non-European Union $70/Members $56
p&p European Unoion £5.00/Non-EU $10.00 per order

Orders to: IOM Communications Ltd, Shelton House, Stoke Road, Shelton, Stoke-on-Trent ST4 2DR
Tel: +44 (0) 1782 202 116   Fax: +44 (0) 1782 202 421   Email: Marketing@materials.org.uk

N. American orders to: Ashgate Publishing Co., Old Post Road, Brookfield, VT 05036, USA
Tel: 802 276 3162   Fax: 802 276 3837   Email: info@ashgate.com

IOM Communications
Reg. Charity No. 1059475   VAT Registration No. GB 649 1646 11
IOM Communications Ltd is a wholly-owned subsidiary of The Institute of Materials

MAGNETISM AND MAGNETIC MATERIALS

SECOND EDITION

J. P. Jakubovics

Magnetic materials have many interesting and important uses. This book discusses the subject of magnetism and magnetic materials at a level suitable for undergraduates studying Materials Science, Physics or Engineering. It gives a brief introduction to the fundamental ideas in magnetism, and then explains the classification of materials according to their magnetic properties. Each of the main classes of materials is discussed in detail. The book then concentrates on materials that are useful for practical applications, namely, those that are strongly magnetic. The properties which determine the usefulness of these materials are discussed, as well as methods of determining these properties experimentally. The final chapter describes materials according to their applications, such as materials for transformer cores or permanent magnets, and it concludes with a discussion of magnetic materials used in information storage. The book includes a set of worked examples, and a set of test questions.

Book 573   ISBN 0 901716 54 5   Paperback
European Union £25 / Members £20
Non-European Union $50 / Members $40
p&p European Union £5.00 / Non-EU $10.00

Orders to: IOM Communications Ltd, Shelton House, Stoke Road, Shelton, Stoke-on-Trent ST4 2DR
Tel: +44 (0) 1782 202 116   Fax: +44 (0) 1782 202 421   Email: Marketing@materials.org.uk

N. American orders to: Ashgate Publishing Co., Old Post Road, Brookfield, VT 05036, USA
Tel: 802 276 3162   Fax: 802 276 3837   Email: info@ashgate.com

Credit cards accpeted:

IOM Communications
Reg. Charity No. 1059475   VAT Registration No. GB 649 1646 11
IOM Communications Ltd is a wholly-owned subsidary of The Institute of Materials

INTRODUCTION TO
CREEP

R. W. Evans and B. Wilshire

Creep and creep fracture represent one of the major problem areas associated with the selection and use of engineering materials for high temperature applications. This text provides a comprehensive introduction to the subject, aimed at final year undergraduate and taught masters programmes, as well as graduates and researchers new to this important field. The coverage includes techniques for measuring and analysing creep properties, the mechanisms of creep and creep fracture, the development of high-temperature materials and the factors influencing their service performance.

The text is copiously illustrated and is supported by an appendix of informative worked examples.

Book 429   ISBN 0 901462 64 0   Paperback
European Union £15/Members £12
Non-European Union $30/Members $24
p&p European Union £5.00/Non-EU $10.00

Orders to: IOM Communications Ltd, Shelton House, Stoke Road, Shelton, Stoke-on-Trent ST4 2DR
Tel: +44 (0) 1782 202 116   Fax: +44 (0) 1782 202 421   Email: Marketing@materials.org.uk

N. American orders to: Ashgate Publishing Co., Old Post Road, Brookfield, VT 05036, USA
Tel: 802 276 3162   Fax: 802 276 3837   Email: info@ashgate.com

Credit cards accepted:

IOM Communications

Reg. Charity No. 1059475   VAT Registration No. GB 649 1646 11
IOM Communications Ltd is a wholly-owned subsidiary of The Institute of Materials